# Defending against Climate Risk

# Defending against Climate Risk:

## *Lessons and Stories from a Foot Soldier in the Climate Wars*

By

Gary Yohe

**Cambridge**
**Scholars**
Publishing

Defending against Climate Risk:
Lessons and Stories from a Foot Soldier in the Climate Wars

By Gary Yohe

This book first published 2023

Cambridge Scholars Publishing

Lady Stephenson Library, Newcastle upon Tyne, NE6 2PA, UK

British Library Cataloguing in Publication Data
A catalogue record for this book is available from the British Library

ISBN (10): 1-5275-9357-6
ISBN (13): 978-1-5275-9357-2

To my wife, Linda, whom I will love forever. You made it all possible with your belief in my commitment to fighting climate change and your support of my efforts.

To my daughters, Marielle and Courtney, whom I also loved from minute one and for whom an earlier, less intelligible draft of this book was a Christmas present in 2018.

Finally, and most importantly, to my granddaughters. Linda and I were there within 8 hours of your birth. Katie and Carrie Madigan, I wish that my work combined with the efforts of thousands of scientists from around the world (many of whom I know) would have left you a healthier world and a more robust planet to inherit. It was not for lack of trying.

I have enormous faith that you will do better.

# TABLE OF CONTENTS

THE ASTERISKS ARE INTENDED TO LEAD YOU TO THE STORIES –
MORE ASTERISKS MEAN MORE STORIES

THE REAL STORIES OF MY CLIMATE LIFE ADVENTURES
ARE IN CHAPTER 13

# LIST OF ILLUSTRATIONS

# LIST OF TABLES

# LIST OF ABBREVIATIONS

| | |
|---|---|
| ACC | America's Climate Choices from the NAS |
| CDC | Centers for Disease Control (and Prevention) |
| CFC | Chlorofluorocarbon |
| CNN | Cable News Network |
| $CO_2$ | Carbon dioxide |
| COP | Conference of the Parties (of the UNFCCC) |
| COVID or COVID-19 | The novel coronavirus |
| EPA | Environmental Protection Agency |
| EPRI | Electric Power Research Institute |
| EU | European Union |
| FDR | Franklin Delano Roosevelt |
| GHG | Greenhouse gas(es) |
| IPCC | Intergovernmental Panel on Climate Change |
| MIT | Massachusetts Institute of Technology |
| NAS | National Academy of Science |
| NASEM | National Academies of Science, Engineering and Medicine |
| NCA | National Climate Assessment |
| NIH | National Institutes of Health |
| NOAA | National Oceanographic and Atmospheric Administration |
| NPCC | New York (City) Panel on Climate Change |
| NRC | National Research Council (U.S.) |
| OSTP | Office of Science and Technology Policy |
| RFC | Reasons for Concern |
| SCC | Social cost of carbon (or carbon dioxide) |
| STEM | Science, Technology, Engineering, and Mathematics |
| U.S. | United States of America |
| UNFCCC | United Nations Framework Convention on Climate Change |
| USGS | United States Geological Survey |
| WCRP | World Climate Research Program |
| WGI or WGII or WGIII | Working group I, II, or III in an IPCC Assessment |

# PREFACE**

Why did I write about the climate "wars"? Why did I think that I was a foot soldier? Why was my life an attack on climate change?

I wrote this book because many people were, and are still, dying with increasing frequency in "projectable" floods, storms, fires, heat waves, and other extreme events all around the world. It makes me think that we were not sufficiently persuasive. Other people, like Steve Schneider who died on an airplane on one last trip to try to maintain defenses against people who would rather make things up for their own wellbeing than save some unknown person's life, whether that be today or sometime in the future.

Failing in the ability to defend themselves publicly against Steve and others, these opponents of climate action would make threats online and in other media against anybody who appeared on their radar screens. Precise details will not be forthcoming, but we are the foot soldiers.

Ben Santer endured the threats and the kidnapping of his son; he is a high-ranking officer in the climate change army.

Michael Mann never blinked.

For me, though, Stephen Schneider was the general.

I am not in the same category as these people, but I have been a soldier in their army. I am still alive, and opponents have failed to damage me and what I have been writing for more than 40 years. They have also failed to damage my family, though they have threatened to do so.

I think that the lessons that follow are important. They reflect what I learned and contributed to the body of knowledge that supports the need for climate action to be taken today. The stories are more fun, though. They reflect what I did, with whom I did these things, and where I was.

I think that I contributed to the common global good, and I hope that you will think that such a view is not a delusion on my part. I had a good time through thousands of experiences with hundreds of people who are now good friends, colleagues, and collaborators from six (maybe seven) continents.

Most importantly, I am still around for my granddaughters. They ask, "Papa, what did you do? What are you doing now?" I think that this book is a pretty good answer to the first question. I am still working on my answer to the second.

What could be better than that? I am alive to know them. They are extraordinarily engaged at a young age, and I can tell them that I tried.

What follows are memories, as well as highlights, of my contributions to the greater good, organized in chapters of unequal length.

# CHAPTER 1

# BASIC TRAINING –
# THE VALUE OF A LIBERAL ARTS EDUCATION

*The lede here is that the value of a liberal education cannot be overstated.*

*Speaking to climate colleagues from many disciplines has been a challenge for me, spanning over more than 45 years since graduate school, but knowing a little bit of their vocabulary and recognizing that their perspectives are just as valid as mine turned out to be very important.*

This first chapter is ultimately about the academic freedom that allowed me to "follow my nose" from the "other side of the desk", leading me from economic theory to climate change. That is an enormous gift from Wesleyan University and my family, but I was prepared for that future by accidents that happened during my undergraduate and graduate educations.

I learned the vocabulary and the value of scholarship from wherever it came from in my random course selections at Penn. I majored in English, then Philosophy, then Chemistry, then Chemical Engineering, and then Mathematics, with a passing grade at graduate level Physics (where the lowest possible grade for an undergraduate was a C).

Some of my decisions were based on my participation in D-1 athletics; I played varsity golf for all four years. Some of my decisions were based on looking forward to what I might be doing at age 40; I did not know then that I would also be doing the same thing when I was 70+ years old. Why? Because it is still fun. Here are some of the highlights:

- My roommate Fred (Sanfilippo now at the School of Public Health at Emory University) and I were taking organic chemistry together during our sophomore years. We worked together on everything but exams. We were separated in exams because I learned so much from Fred's exam prep sessions that they thought we were cheating when we both blew out the curve on the first two exams. The truth is, we had taught ourselves enough that all we needed was access to the periodic table of elements. That table was always hanging on the classroom wall. Since we could see the table, Fred and I thought that we could cope with just about anything that would be thrown at us on an exam - and we were right. Persistent grades above 95 percent were not expected from anybody when the median was around 50.

I learned from our study technique. As we studied, Fred would suggest some patterns in the chemical equations of interest, and I would check their implications against our notes and the textbook. Sometimes Fred's magic worked, and sometimes we would argue. By the time we took the exam, we had discovered maybe 5 or 6 patterns that were the truth, and that was all we had to remember (as long as they did not put a sheet over the periodic table – then we had to remember more).

The Chemistry Department was so sad that neither of us wanted to major in chemistry.

- I played intercollegiate golf (D-1) during my four years at Penn. I qualified for the NCAA tournament in 1968, but did not make the cut. In my junior year, in the spring of 1969, we had 22 away matches in the month of April. We traveled away from Philadelphia so much that I was on campus for only one day the entire month.

This was before the internet, but fax technology worked so that I could get my assignments to my very gracious professors on time. I kept up in five courses (even though the normal load was 4) and earned 5 A's.

I was elected to Phi Beta Kappa in my senior year - to my surprise and despite my interest in golf. It was the 1960's, so the rules were different. But the PBK rules were about performance. To be honest, I did not even know about PBK until I got their invitation letter.

- I learned some important lessons from commuter train rides back and forth from my parents' home at the end of the "Main Line" during my years at Penn. There were so many sad faces on the morning train from Paoli to Penn Station in Philly on Monday mornings when I traveled back to campus from weekend visits with my parents so that I could spend 12 hours a day on the practice range.

Most of the riders were reading something like National Geographic for relaxation and distraction, but they looked like they were about to be hung – so sad to be going to work that it was unbearable to watch.

I decided then that I did not want that. I did not want to ride that train. I wanted work to be fun. I wanted to be able to say that I never did anything at work that was not fun. Looking back over 50 years, I can say that that objective was accomplished.

- A job in academics seemed like the plan to avoid those train rides. There was, though, difficulty conveying what that meant to my educator parents – a high school principal and a kindergarten teacher. They did not understand that I was not just going to be a teacher. They did not understand that research did not mean going to the library to read what somebody else had already written. I was going to be a researcher for whom the expectation was that I would expand the knowledge frontier of whatever discipline I happened to choose. My task would be to write something new.

They never really got it. When I was an Assistant Professor at SUNY Albany in 1976, I accepted my first invitation from another campus to give a talk about my work. I would be speaking at Lehigh University in Bethlehem PA. I spent the night before the talk with my parents in their home in Hershey's Mill just outside West Chester, PA. When I got up for breakfast, my father joined me. I remember nothing about what we ate, but I do remember his question when I was collecting my things, my thoughts, and my nerve to leave to give a talk about the general equilibrium implications of environmental policy on real relative wages for labor in a steel town. "Why do they want to listen to you for, anyway?" was his question for me as I left the house.

Thanks, dad. I was already nervous enough.

The talk was not well received, but that was not a surprise. What was well received was its rigor in applying techniques from papers by Paul Samuelson. I had not created a new approach, but I had adequately applied some existing ones.

- Having ultimately majored in mathematics, I went to grad school in mathematics at SUNY Stony Brook. I loved math and the isolated worlds that it built. They protected me from the violence across the outside world (MLK and Bobby Kennedy had been shot). I was about to go to work for Bobby's campaign in CA and would have some responsibilities during the Democratic Convention. Then, he was killed.

- Then I just wanted to retreat into my own world, and mathematics was an option.

  I really wanted a job on the other end of graduate school, though. I worked hard for two semester in a graduate math department. I was a good student. But some of my classmates were much better than I. They saw everything intuitively. They did not, it appeared to me, work at all, except for a few hours early Wednesday mornings (listening to Simon and Garfunkel at 3AM?) because assignments were generally due on Thursday. Still, they got everything right and laughed about it. I got it right, but it was difficult.

  I knew that they were going to get jobs as mathematicians wherever, and I learned that I was the academic mechanic who could push through the proof and get the same answer as they one week later. I was pretty sure that I would drive a taxi and they would put men on the moon. And so, I changed majors again – to economics in the middle of graduate school.

- After some advice from the SUNY math department, I switched to economics (with one first semester intro econ course at Penn to my name). I applied to PhD programs in economics (not business) at Penn, Harvard, Princeton and Yale. (Stony Brook had told me that I could just switch fields, so I had a safety school. I was accepted at Penn and Yale.

  Penn offered a full ride. Yale offered no support other than no tuition and a graduate teaching position starting in my second year if I could prove that I could teach undergrads. I chose Yale, even though I did not realize how good the Yale Economics Department was. Turns out that there four Nobel Laureates in the house.

The Yale Economics Department admitted me as part of an experiment crafted by Herb Scarf and Joseph Stiglitz* (an asterisk will henceforth indicate somebody who has won a Nobel Prize in Economics). They pushed the Department to admit to me and Andy Rosenburg – both math major from different schools, but each with minimal economics backgrounds.

When I arrived at Yale, or as I would say on the train "New Haven for graduate work", the economic vocabulary was foreign. It follows that I was ahead in the running for "who, from my entering class, will learn the most economics?" I could teach the math to my peers while they taught me the economics. Willem Buiter and Robert Wilson and I would become the last (as far as we know) to complete Yale PhD program in economics in 4 years.

I am the last by actual count, since Willem and Bob got their degrees minutes before I did (alphabetically, Buiter before Wilson before Yohe).

- But let's take a look at what that education involved. I learned microeconomic theory from Joseph Stiglitz*. I learned macroeconomics from James Tobin*, with copious notes provided by teaching assistant Janet Yellen. I learned mathematical economics from Tjalling Koopmans* and Herbert Scarf. I learned environmental economics from William Nordhaus* and Tjalling Koopmans*. Martin Weitzman would be the inspiration of my dissertation, even if he did say "you can't do that, it is too hard."

When I took my qualifying orals to move into the dissertation stage of my time at Yale, my examination committee included Joseph Stiglitz*, James Tobin*, Richard Cooper and Richard Becker. After my two-hour exam, I waited outside the exam room for their decision. Did I pass????

An hour later, Stiglitz stuck his head out the door and told me to come to his house tomorrow morning (Saturday) at 10 AM. My future wife, Linda, was waiting for me to hear the news. We did not speak much during the walk back to the Hall of Graduate Studies.

A sleepless night later, I arrived the Stiglitz house on Livingston Street to find the future Nobel Laureate aerating his front lawn with a rake. He was happy to stop when he saw me. We sat on his stoop, and he told me that the examination committee had decided that they would not pass me until I learned everything in the Samuelson's* introductory economics textbook. They would reconvene in the fall to examine my understanding of the principle or economics.

Their decision was the right one, but very unprecedented. It seems that I got all the hard questions right, but I got all of the easy ones wrong. "Come back in 4 months and then we will examine you on that material in Samuelson's* intro textbook. Nothing else, but don't forget the footnotes". It turns out that I actually wrote study guides for subsequent editions, but that is another story.

I read and worked through the text carefully over the summer, and I passed easily in the fall.

For somebody without an undergraduate background in economics, it turns out that this was the best thing that ever happened to me. It gave me the skills to teach at Wesleyan, and it made my academic papers better. I got into the practice of teaching my latest academic paper to be submitted to world class journals because, if I could not make it make sense to engaged and intelligent undergraduates, then I did not know what I was talking about. I now work to communicate issues surrounding climate change to lay audiences, but I can do that because what they took the time to teach me. I have learned again that you have to be able to reach back to first principles – and that was the lesson that the members of my examination committee were teaching me.

The names in the last paragraphs were ordinary in my life in New Haven. I saw them every day. Those without asterisks include Richard Cooper, who had recently been Undersecretary of State for Economic Affairs under Jimmy Carter; Herbert Scarf, who brought fixed point theory to economics and should have won the Prize; Martin Weitzmann who would bring the black swan dark tails of outcome distributions to climate change economics; and Janet Yellen, who would become Janet Yellen. ALL of the rest have asterisks next to their names because they would eventually win Nobel Prizes.

Yes… All of them. Nobody knew at the time, but I was at Yale and what did I know. The culmination of my education there was like calling on Mariano Rivera to pitch in the 9th inning of a 1995 post-season game for the Yankees. He was good. He was engaged. He had a spectacular slider. But was not yet "Mariano".

Linda and I married in September of my 4th year. I played in the USGA Amateur Championship at Ridgewood Country Club in New Jersey the week before. I lost in the first round, but I had qualified the two weeks earlier with the lowest score in the highly respected Philadelphia region. I shot 140 for 36 different holes. If anybody is counting, that was 4 under par.

It was not easy to qualify out of Philadelphia, and Linda had a role. I was feeling pressure and I barely bogeyed the 9th hole of the second round. I was leading but this was just the time for a collapse. Linda, who was not allowed to follow me around the course because women were not allowed on the gounds of Philmont Country Club, caught up to me at the tenth tee (which was close to the terrace where she could spend her time without being able to order anything). She knew what had just happened. "Time for a birdie" she hollered when I approached the tenth tee. On a 240 yard par three. Yeah. Right. A birdie picks up two shots on the field. I agreed and smiled. She had broken the tension. I hit a 239 yard one iron within 3 feet of the hole – I scored her birdie 2 with some jittery nerves even though it was a straight-up hill put.

It was easy from there because Linda had given me my confidence back.

Nonetheless, Linda was not amused about the USGA National Amateur Championship that occupied my attention during wedding preparations. I lost in the first round.

She was not, however, about to stand in the way of my completing on something that mattered. Later that year, she insisted that I would complete of dissertation on time (4 years after I entered the Yale program and, in the view of my parents, the minimum even after we were married. She was not going to be blamed for my "failure".

I wrote on a "desk" in our apartment in Albany (where I had taken a job to pay the rent) that was simply a panel door placed on top of boxes of books. Linda protected me from students (who lived nearby since we were living in low rent housing. She supplied coffee, and I worked on the arithmetic support of "prices versus quantities under uncertainty" (that is now known as taxes versus cap-and-trade, but more on that later).

When I had to get my dissertation typed, Linda and I and our two cats stayed (in an infirmary room thanks to Willem Buiter) when we (not I) went to New Haven to consult with William Brainard (my dissertation adviser) and my typist.

Brainard would never admit to reading anything that I had written I preparation for these meetings – one chapter after another. That was part of his teaching style for, I came to understand, is most promising students.

He would make me present my findings in his office with nothing but chalk and a blackboard. I quickly learned that this was going to happen. The subsequent give and take from that practice were some of my best learning days of my life. I was learning that I could play in the Yale-level game (which it turned out was Nobel worthy), but I also learned that I was not always right. I learned about "laugh tests" as in, from Brainard, "that cannot possibly be true – consider this (made up) example".

When each presentation was over and he was not convinced that I was right about a particular claim, I came to know that he was always right to be skeptical and that I had more work to do. I also knew why and made up possibilities were part of the process. I taught the concept of "laugh tests" for nearly 50 years.

That is to say, his ability to direct and dissect my dissertation is something that I carried with me throughout my academic career. Not just for my students. Middle of the night periods of staring at a dark ceiling upon which I could mentally draw graphs and write text became the norm. For years….

Why was this way of living productive? I would always remember what had happened in the dark, so it was in the short run. Did it turn out that these episodes were not healthy? Yes.

To support these mid-night episodes in the short to medium run, it turned out that my accidental multidisciplinary background at the University of Pennsylvania would support my interdisciplinary work in climate change with natural and physical scientists, as well as other social scientists. I could talk to them because I knew a little bit about lots of things. I could write with them because I knew some of their vocabulary and they would fix my mistakes. Ultimately, those eight years playing very competitive golf in a protected environment became the foundation of my professional life.

A few years after my last national amateur and also after spending a few years writing significant insights into the economics of decision-making under uncertainty (as indicated by where my papers were published), I found a lasting home at Wesleyan University. I was hot stuff at but time, but I wanted more (or less); and I was not sure if I was their ideal candidate.

Wesleyan was searching widely, but I had applied to one school – not because they were the only school looking for my demonstrated skills, but because they applied consistently to their liberal education philosophy of equally across disciplines for the student body and for the faculty. And also because I had met my wife in Connecticut.

As I am sure that they expected, I stopped publishing in economics journals late in the 1980's. I started to publish a lot in climate and science journals – sometimes big deals like *Nature* or *Science*. Sometime *Climatic Change* that was launched with some professional risk by a future friends and menor – Steve Schneider. That was OK with my colleagues, even though I had come to them as an economic theorist. They were happy that I was making a contribution to the public welfare in journals that they could retrieve and read.

For me, their evaluation of my value to the University was the gift of a lifetime. The rest of this book covers the consequences of their decision.

I hereby thank them and the University for this freedom, but it was not unexpected. Wesleyan is, after all, the place where "academic freedom" was invented.

Later into my tenure at Wesleyan, I was invited to give a 12-minute talk to the Board of Trustees (the Chair was an attorney, so bill-able hours were measured in tenths of an hour). He wanted me to speak to tenure decisions for junior faculty working inter-disciplinarily.

I had published 25 papers in the previous five years (none in economics, per se, but all have many citations – some as high as 12,500). I argued that a positive tenure decision on the basis of that record would have been appropriate at Wesleyan, but that it would have been impossible in a standard economics department at places like Yale or Michigan or Stanford.

Wesleyan did not agree to that standardized code. Anywhere else? I looked, and the answer from a limited sample was no – at least not for a typical top-tier Economics Department.

# CHAPTER 2

# PRICES VERSUS QUANTITIES
# UNDER UNCERTAINTY

*The lede here is that cap-and-trade markets for allocated permits are always preferred to fixed standards for all sources of emissions of a pollutant. They may not, however, be preferred to a price (a tax) control, depending on the variability in total emissions and the resulting losses in expected benefits on the demand side of the product markets.*

My PhD dissertation, *A comparison of prices controls and quantity controls under uncertainty*, is a microeconomic theoretic exploration of questions born from a seminal paper entitled "Prices vs. Quantities", authored by Martin (Marty to his friends and enemies around he world) Weitzman in the *Review of Economic Studies* in 1974.[1]

Marty's paper was perhaps the first of many that showed us all his incredible skill in framing complicated questions as simple, analytically tractable propositions whose explorations and explanations would nonetheless illuminate the intricacies of the larger motivating context.

Here, he assumed a single firm facing a single market with quadratic benefit (profit) and cost functions. The firm knew its cost schedule very well, but demand was variable and not predictable from one time period to the next – sometimes it would be high, other times low, and occasionally average.

There were more subtleties than that behind his paper, of course, but I ran with the fundamentals - trying to understand the intuition behind the simple case before tackling more complicated possibilities of multiple firms.

Starting simply allowed me to understand that the important questions in the decision-making under uncertainty are "Who knows what? When do they know it? How do they respond?" This is where I came to understand that the answer to every economic question of any economic consequence

---

[1] https://scholar.harvard.edu/weitzman/publications/prices-vs-quantities

is "it depends". It follows that the real question that we should all be asiing is "on what?" Over the next fifty years, my students would hear that lesson over and over.

I would publish a number of papers on more complicated versions of this prices-versus-quantities comparison in big-deal economics journals after coming out of graduate school (numbers 1, 3, 4, 6 and 8 as well as the dissertation itself, #7; references and links below).

The equations that I worked through provided some insight into the "On what?" question. The equations said that:

- the significance of the choice depends upon the variance of total output under a price control (as opposed to strict quantity standard);
- the direction of the significance depends on the difference between the slopes of the marginal benefit and marginal cost curves; and
- the variance of total output under the price control depends on the slope of the marginal cost curves.

The intuition behind these results turned out to be fairly simple to explain for the single firm model. There were two cases that make this clear:

- Given a quantity control, a single firm would produce up to the specified quantity regardless of the market clearing price. The price would be high for high-demand periods, low for low-demand periods, and average for circumstances close to the mean. The benefits of quantity restrictions would stay the same, given that the specified fixed quantity would clear the market regardless of demand. As well, the single firm market would be unique and isolated by Weitzman's assumptions, so there is no place for secondary changes in benefits.

- Allowing output to vary depending on demand conditions by setting a price control for which the *expected* output matched the quantity standard would allow the firm to increase its expected profits. Compared to the average, output would climb for high-demand and fall for low-demand – both to the benefit of the supplier. How do we know that will happen? Because the supplying firm would not change its output if it were not worthwhile to do so.

Unfortunately, variable output allowed by a price control decreases expected benefits for ordinary citizens to an extent determined by the curvature of the consumers' benefit curve. The reason is that increases in quantity above the average increase benefits society more slowly than reductions below the average cause harm.[2]

It follows, that we had discovered a tradeoff. Would you, if you were to move from a standard to a price control that would achieve the same outcome on average, achieve efficiency gains to the firm that would exceed the damage done to consumers? Maybe, but maybe not. And why not?

Here are the two cases so you can consider how applications of this intuition can inform action decisions for two pollution examples:

*Case 1:*

An emitting firm would always release up to the allowable quantity under the quantity control (and maybe more if the fine for violation were small).

*Case 2:*

Given a price control, the emitting firm would vary its emissions. It would emit more when demand for its product was high, and emit less if demand were low, and medium emissions when demand approximated the average.

We can now try out the\is intuition for these two specific examples – still for a single firm and a single market.

First of all, for example, consider sulfur emissions. In this case, annual emissions matter and there exist existential thresholds. It follows that annual variability in increased emissions can do extraordinary harm. Damages could go up more quickly during periods of high product demand that generate high emissions, but they would fall more slowly when demand is lagging.

Flexibility in emissions allowed by a tax would, therefore, be potentially extremely expensive in terms of the economic accounting of environmental

---

[2] This is a reflection of diminishing marginal utility – a fundamental assumption in most of economics. A little more is better, but not so much if you are rich, and really a lot if you are poor.

damage, but this extra damage could be avoided by setting a quantity standard.

For carbon emissions, though, damages depend on temperature increases which themselves depend on *cumulative emissions*. It follows, therefore, that annual variability in emissions around a predictable annual average does not add to expected costs as long as cumulative totals over a specified, relatively long-term time horizon are constrained. Emissions may look large in any given year, but they would be smaller in other years.

It is here, based on straight up economics, that an emissions tax (this time on carbon) would be preferred, because the variability in emissions is essentially harmless over time.

Things get a little more complicated with multiple sources of pollution. The tradeoff still hinges on the variability of cumulative emissions, but now it is the sum of multiple firms' collective actions.

With a price control, they all face the same price for each unit of pollution emitted and they could bargain to make things better – for themselves and for society.

With firm specific quantity controls, though, total emissions would be fixed because all firms' emissions would fixed. But would that be optimal

Within a cap-and-trade environment, firms could buy or sell permits from each other so that they could respond to high or low demand in their own markets as much as they want – that is, *they can maximize their profits subject to the constraint that their net total activity in the emissions market would cancel out – and therefore cause no additional harm.* In that way, lost expected benefits from pollution variation would be eliminated.

Simple application of this intuition confirms that a cap-and-trade regulation always dominates setting strict and firm specific standards for every emitter. Total emissions are fixed under both, but cap-and-trade regulation allows some flexibility across firms that makes them more profitable (otherwise, they would opt out of trading). The environment does not care where the pollution comes from, so the key here is that buying or selling permits will only occur if it is in two firms' best interest – meaning that net profits will increase.

Nonetheless, the Weitzman tradeoff still applies in the aggregate choice between a price (a tax) and a *total* quantity constraint with a permit market.

In the environmental world, this policy tradeoff still comes down to the extra damage caused by the sum of variable total activity under the tax. To see how, return to thinking about total carbon emissions and/or total sulfur emissions through the lens of variable emissions from year to year.

Since it is cumulative carbon emissions that cause temperatures to rise and cause thereby cause damages, variation from year to year need not add expected cost even from many sources as long as their total emissions average to levels that match the standard for total emissions.

For sulfur emissions, though, there are damage thresholds for each year, regardless of the number of emitters. Going above these thresholds in any year can produce enormous extra cost, regardless of any value from a permit market. It follows that quantity controls on sulfur emissions per year (even with a cap-and-trade program within an air-shed) is the better policy approach. Why? Because total emissions from all sources are constrained below a critical threshold.

It should be clear, then, that a price (a tax) would the better choice for carbon emissions - as long as it changes over time to track a least cost emissions trajectory.

Richard Schmalensee (an MIT economics professor and member of the Council of Economic Advisors under George Herbert Walker Bush) and I had a conversation about this conclusion one morning in 2009. We were on a morning bus ride going to an America's Climate Choices meeting at the National Academy of Sciences. He had "been in the room" when the approach to sulfur emissions targets were decided.

He agreed with my economics-based conclusion, but he pointed out one critical reality of the US political economy: it turns out that only the House of Representatives can impose a tax or change a tax. They did not respond very quickly, they were allergic to more taxes, and many did not take climate change seriously.

Naturally, therefore, he argued convincingly, that imposing a carbon tax was not viable. Something like a cap-and-trade with maximum flexibility across emission sources from year to year, defined over time by a cumulative emissions constraint that would become more restrictive (think REGI in New England), would be a better idea. It would minimize economic costs, subject to a political constraint, and it would inspire innovation in alternative energy, as well as its marketing. It might be a second-best option according to theory, but it would be a much a preferred choice in the real world, even for carbon, to nothing.

How so? Because the Supreme Court of the US had decided that carbon dioxide is a pollutant. It followed immediately that the Environmental Protection Agency could restrict cumulative carbon emissions however it wanted (with justification through public review about public harm but without permission from the Congress – they already had permission from the Clean Air Act of 1970). In short, the Clean Air Act applied. As I write in January of 2023, though, all bets are off.

Like my father used to say, "You will learn something every day if you're not careful." My father also used to say that you were a "damned fool" if you make the same mistake twice. Dick Schmalensee taught me a lesson about the political economy that I have never forgotten; but ongoing events are frequently troubling.

## References with links and abstracts connected to my January 2023 CV numbers

**#1** Yohe, G., "Substitution and the Control of Pollution--A Comparison of Effluent Charges and Quantity Standards Under Uncertainty," *Journal of Environmental Economics and Management* **3**: 312-324, 1976.
Single-valued price and quantity controls of a polluting activity are compared under uncertainty. The ability to substitute other inputs for the pollutant in the production of a positively valued final good, and the usual discrepancy between the amount of pollution actually produced and the amount emitted are carefully incorporated. The first is found to influence the degree to which cost fluctuation is reflected in the output of the final good. The second concern alters the region of the benefit function into which output is inserted. Both change the welfare losses are associated with random fluctuation in the costs of reducing pollution.

**#2 Yohe, G.**, "Polluters' Profits and Political Responses: Direct Control versus Taxes", *American Economic Review* **66**: 981-982., 1976.
In a recent issue of this Review, James Buchanan and Gordon Tullock (B-T) sought to present a positive theory in explanation of the frequency with which direct controls of an externality are imposed in lieu of punitive taxation. They argue that this frequency is observed despite the preference of most economists for price controls, because those economic actors whose production or consumption is to be regulated not only prefer direct quantity control, but also possess the means with which to press their will upon the political decision maker. It has been the point of recent work in the theory of regulation under uncertainty that the economists' general preference is not entirely well- founded (see for example Marc Roberts and Michael Spence, Martin Weitzman, and the author). There do exist many quite plausible

situations in which both economists and the Buchanan-Tullock regulatees should prefer quotas

**#3** Yohe, G., "Single-Valued Control of a Cartel Under Uncertainty--A Multifirm Comparison of Prices and Quantities" *Bell Journal of Economics* **8**: 97-111, 1977.
Homogeneous and hybrid price and quantity controls of a cartel seeking to maximize cumulative profits are compared within an uncertain economic environment. The primary determinan1t of the superior control is shown to be the relative influence each choice has on the variation in total output. A member firm's size, relative to the total output, and the correlation of its output with the outputs of the other firms are therefore crucial in predicting whether the firm should optimally face a price or a quantity. Extensions of the analysis to pollution control, agricultural supports, and planned economies are also outlined.

**#4** Yohe, G., "Single-valued Control of an Intermediate Good Under Uncertainty--A Comparison of Prices and Quantities" *International Economic Review* **18**: 117-133, 1977.
The present study is such a construction; we will compare once and for all price and quantity control of an intermediate good that derives its value entirely from the final good it is used to produced. Careful attention will be paid to the impact of the elasticity of substitution in the production of the final good and to the profitability of producing that good in the face of either type of regulation. While it is a tribute to the versatility of the Weitzman framework that such structure can be successfully incorporated, it will become apparent that his basic model falls well short of handling the intricacies of our problem.

**#6** Yohe, G., "Towards a General Comparison of Price Controls and Quantity Controls Under Uncertainty" *Review of Economic Studies* **45**: 229-238, 1978.
In a recent article published in this Review, Professor Martin Weitzman [8] developed a cost-benefit model designed to explore the general belief held by most western economists that price controls are a more efficient means of regulation than the corresponding quantity controls. Weitzman effectively concentrated his attention on this preference by analyzing only the second best choice between singular, once and for all controls of a production activity. Our present purpose is to extend the Weitzman analysis of the single firm case to include additional sources of uncertainty and informational difficulty that might naturally appear within a regulated hierarchy. In so doing, we suggest a slightly different, but more accurate interpretation of the results that can free us from the assumption that the regulated firm be a profit maximizer.

**#7** Yohe, G., *A Comparison of Price Controls and Quantity Controls Under Uncertainty*, New York, Garland Publishing Co., Inc., 1979. (Published in a series of 24 Outstanding Dissertations in Economics).
The prices vs quantities comparison is conducted within a cost-benefit model in which the output of a profit maximizing enterprise is to be controlled. The regulator is constrained to the issuance of either one price order or one quantity order,

determining that control specification by maximizing expected benefits minus expected costs. Uncertainty is introduced1from three separate sources: imprecise knowledge of the cost and benefit functions themselves, the possibility that a quantity order from the center will not be filled exactly, and the chance that the quantity consumed need not equal the quantity actually produced. Any pollution example provides motivation for the final source of uncertainty. The vehicle of comparison is the comparative advantage of prices over quantities is defined to be the expected value of the algebraic difference between the level of benefits minus costs achieved with price controls and the corresponding level achieved with quantity controls.

**#8** Karp, G. and Yohe, G., "The Optimal Linear Alternative to Price and Quantity Controls in the Multifirm Case" *Journal of Comparative Economics* **3**: 56-65, 1979.
Linear control schedules in output have been shown superior in the one-firm case to either of the extreme controls -price or quantity; they optimally trade off the desirable characteristics of both extremes. When many firms are regulated, however, that superiority fades. Then total output affects expected benefits and can display a larger (or smaller) variance than the sum of individual firms' output variances (upon which expected costs depend) if costs are positively (negatively) correlated. Output variation must be discouraged (encouraged), therefore, and the linear schedule rotates toward the quantity (price) extreme. The better extreme might thereby become the best choice among all three alternatives.

**#16** Yohe, G., and MacAvoy, P., "The Practical Advantages of Tax Over Regulatory Policies in the Control of Industrial Pollution", *Economics Letters* **25**: 177-182, 1987.
A tax cum subsidy pollution control mechanism is proposed to mitigate against the potential efficiency losses caused by moral hazard in a self-reporting method that prescribes a best available technology and trusts, in the absence of expected cost penalties, that it will be employed fully.

**#17** Yohe, G., "More on the Properties of a Tax Cum Subsidy Pollution Control Strategy," *Economic Letters* **31**: 193-198, 1989.
A modification in a tax cut subsidy pollution control strategy proposed by Yohe and MacAvoy to mitigate against the effects of moral hazard on the effectiveness of self-reporting strategies is shown capable of eliminating the dead weight welfare loss of regulating the emissions of an imperfectly competitive polluter.

**#24** Yohe, G., "Carbon Emissions Taxes: Their Comparative Advantage under Uncertainty," *Annual Review of Energy* **17**: 301-326, 1992.
This paper reviews a growing literature on carbon taxes - taxes that have been proposed as one policy option that would be available to the international community if it were to choose to slow the rate of greenhouse-induced global warming by curbing the worldwide emission of carbon dioxide. It compares and contrasts published answers to a wide range of important questions. What levels of taxation would, for example, be required to achieve specific emissions reductions?

How much aggregate economic activity would be sacrificed as a result? How would these taxes and their associated efficiency losses evolve over time if the specified emissions reduction were to be permanent? How would the answers to these and other questions be altered by adopting alternative baseline scenarios of what an unregulated future might hold? How sensitive are these answers to alternative assumptions about where and how the enormous tax revenues might be distributed and then spent?

The ability to adapt to change and to cope with more severe extremes would, however, be linked inexorably to the second set of socio–political–economic scenarios. The second dimension, defined as "anthropogenic" social/economic/political scenarios, describe a holistic environment within which the determinants of adaptive capacity for water, agriculture, or coastal zone management must be assessed.

# CHAPTER 3

# VERY EARLY CARBON EMISSIONS AND CONCENTRATION TRAJECTORIES

*The lede here is that William Nordhaus got me involved in climate change issues by inviting me to participate in a National Academy of Sciences study in 1982. It turned out that we would set a standard for modeling given multiple sources of uncertainty. It also turned out that it would change the course of my professional and personal lives forever. References to my contributions, abstracts, and active links to my January 2023 CV are appended at the end of this chapter.*

As a member of a panel of the National Academy of Sciences (NAS then but now the National Academies of Sciences, Engineering and Medicine NASEM) that was commissioned to produce a report on *Changing Climate,* William Nordhaus offered to project a range of emissions scenarios for carbon or carbon dioxide through 2100. The panel agreed that such an effort would be worthwhile even in the early 1980s. Unlike many Academy panels, this was actually a project that allowed its panel members to do new science, and Bill was game.

Bill called me out of the blue on a winter's day in 1980 to help. He asked if I would be interested in helping to produce and interpret a wide range of emissions scenarios. It seemed like a good idea to me at the time, even if I did not know anything about global warming or its drivers.

Bill was my mentor from Yale, and he promised that he would fill in the gaps. After all, what could there be if we did not invent it. There may have been 5 economists on the planet at that point in time who knew anything about global warming and climate change. I agreed enthusiastically, and the rest is history.

When we started our work, we took our task to mean that we should do more than produce ranges of emissions and concentration scenarios from a single economic model. We thought that we should also investigate, even within a credible aggregate global economic model of our own construction, the degree to which each of multiple sources of quantifiable uncertainties influenced the variance emissions most significantly. Our chapter in the ultimate NAS report is available as #10 in the papers collected below.

Before I get to the details, it is important that I share a personal note. Linda and I bought a Yamaha piano for our children with my fee as a consultant to this study. This payment was just about the only external money that I have ever received as payment for anything that I have done outside of Wesleyan or any other academic institution over nearly 50 years of work on climate (except for some modest 1990s EPA funding for research on sea level rise and some occasional consulting work on adaptation).

I had *never* before been compensated for contributions to global science Nor would I ever be compensated for my time devoted to assessments through: the Intergovernmental Panel on Climate Change, various National Climate Projects, Michael Bloomberg's Risky Business analysis, the New York Panel on Climate Change, National Academy work to produce decadal for research strategies for NASA, and any other National Academy of Sciences studies including its Stabilization Panel and the Adaptation Panel and the Overarching Committee Report on America's Climate Choices. – and so on.

I may not have been compensated, over nearly a half century for extra-curricular activity, but I can confirm that I always got more out of my gratis experiences than I put in - above and beyond coverage for travel expenses and meals which was more than enough. It has been a life's experience, and thus so many more chapters.

I added up the dollar value once through the summer of 2016. As of then and using my then-current private consulting fee of $200 per hour, the time that I had donated *gratis* to publicly motivated assessment efforts for the United States and the planet totaled more than $2 million. For me, it was all worth the sacrifice. For Linda, I am not so sure given all of the travel. For skeptics, though, it did not include the value of all of my private jet travel to distant meetings – never happened.

Returning to the *Changing Climate* work, Bill and I produced spaghetti graphs, presented here in Figure 3-1, by constructing an aggregate economic model of global production driven by labor, fossil fuel, and non-fossil fuel plugged into the best climate model of the time. Some reduced form description of the climate science model was required (not without some internal discussion and dissention across the panel) so that fossil fuel consumption that produced carbon emissions would also be seen to produce projections of atmospheric concentrations of carbon dioxide through 2100.

A point of reference is essential, here. Even in the early1980s, it was widely known that carbon dioxide was a greenhouse gas that was capable of warming the planet.[3]

Our synthetic modeling included ten major sources of uncertainty, so our background required that we produce distributions for each. The left column of Table 3-1 lists them holding each constant. The right hand column shows rankings assuming that everything else was fixed at its median. The airborne fraction ranked sixth in both lists.

Doing the modeling turned out to be tricky, though, because Uzawa (1962) had proven a theorem that seriously diminished the applicability of production schedules with constant elasticities of substitution in cases for which more than two inputs must be considered.[4] His finding states that if the elasticities of substitution between every pair of inputs are to be held constant (which we needed to specify the model), then one of the following two conditions must be satisfied:

(1) the elasticities of substitution between all input pairs must be identical, or
(2) the elasticity between at least one pair of inputs must be equal to -1.

Neither assumption would work for us, so it was one of my tasks to develop an approximation procedure that would be rigorously consistent with the second Uzawa condition in any year, but change every year so that the elasticity for that pair over time would not be in unity.

---

[3] **Yohe, G.**, 2022, "A Nobel Prize in Physics that we can all understand", *World Financial Review,* July/August, https://worldfinancialreview.com/a-nobel-prize-in-physics-that-we-can-all-understand/
[4] Uzawa, H., 1962, "Production functions with constant elasticities of substitution", *Review of Economic Studies* 29, 291-299.

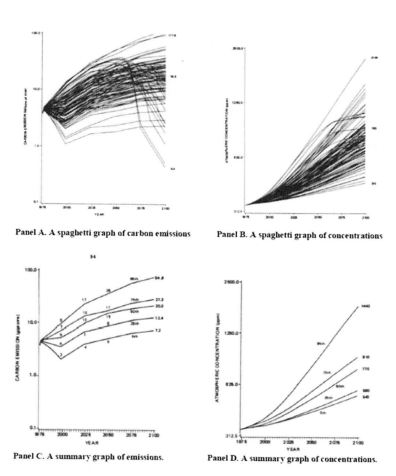

**Figure 3-1. Results from the NAS analysis.** Panels A and B show the raw data for 100 model runs (blurry from the poor quality of the original figures); Panels C and D show summary statistics for emissions and concentration trajectories. Source #10.

**Table 3-1 Indices of sensitivity of atmospheric concentrations in 210 to uncertain about key parameters (100 represents the maximum effect; all others are proportional).**

| Parameter | Marginal variance from full sample [5] | Marginal variance from most likely [6] |
|---|---|---|
| 1. Fossil/non-fossil fuel substitution | 100 | 100 |
| 2. General productivity growth | 76 | 79 |
| 3. Energy/labor substitution | 56 | 70 |
| 4. Fossil fuel extraction cost | 50 | 56 |
| 5. Trends in energy production | 48 | 73 |
| *6. Airborne fraction* | *44* | *62* |
| 7. Fossil fuel mix | 31 | 24 |
| 8. Population growth | 22 | 36 |
| 9. Relative cost energy | (-3) | 21 |
| 10. Total resources of fossil fuel | (-50) | 5 |

Coming out of the weeds for the moment, this entire experience led to playing in "scenario-land" and worrying (privately at first) about how a decision maker might try to cope with too much information. Nobody can cope with hundreds or thousands of "not implausible" scenarios, but everybody should try to cope with an image of the entire distribution of possible futures – good extremes and bad extremes, as well as all the stuff in the middle. I worked on how, statistically, to define representative scenarios in paper #21. I also invented the notion of "not-implausible" futures in papers #41 and #52. Specifically, this led to a thought collaboration with Steve Schneider about how to describe, and what to do within, the damaging tails of "not-implausible" futures in an economic context. More on that later, because it took some time to work on that problem. The answer would appear in 2007.

Returning to the Academy committee chaired by William Nierenberg, I learned another lesson about my standing. We all prepared and released a National Academy report named *Changing Climate* in 1982. It was important for all of us, but it was more important to some - included come on the committee who were the leaders of two research groups with competing estimates of the "airborne fraction" (the fraction of a ton of

---

[5] Variance when the listed variable assumes its most like outcome and the others all reflect their full ranges.
[6] Variance when the listed variable assumes its full range and all others are set equal to their most likely outcome.

emissions that remains in the atmosphere after one year and then persists with a half-life of about 100 years).

This parameter was one of ten sources of uncertainty that Bill Nordhaus and I had included in our planning process and calibrated in our modeling. We could not favor one over the other, and so we included a wide bi-modal range to reflect scientific disagreement about the airborne fraction – one modal estimate below 50% and the other well above 50%.

Years later even in 2023, our work still defines the standard for rigorous emissions modeling, and our spaghetti graphs are familiar not just to climate scientists. Most citizens who get their weather forecasts from television news have seen meteorologists use something very similar to our spaghetti graphs in their coverage of where hurricanes might go, for example.

In between, when it was time to present the results about the airborne fraction to the Committee, things became less clear. The two Bills (Nordhaus and Nierenberg) taught me a lesson. The presentation of Table 3-1 was assigned to me – the rookie in the room – in one of our last sessions and just before dinner.

I explained our method and displayed our results. The room erupted in argument when I finished, and the resulting chaos that could have lasted for hours except for dinner reservations at the Watergate. In fact, the debate would spill over into the next morning (but it did not disrupt dinner). While the initial chaos was ongoing, I looked over to Bill Nordhaus when nobody was yelling at me; he was leaning back against the wall on his chair, laughing and smiling. I looked at Bill Nierenberg, the Chair of the Committee for help, and he just smiled. I took control of the room by adding the debate as the first item on the agenda for tomorrow's morning meeting – the last of our drafting meetings.

"Welcome to the big-time," Nierenerg said later while we adjourned for dinner. "Know your audience," was his advice. He then asked "How are you going to start tomorrow morning's meeting?" By returning to the table was my response. Nordhaus would say the same thing minutes later and he would ask the very same question.

I collected my thoughts and framed my answer questions that I now anticipated. We will stand by our work - given our results not withstanding your disagreement, you will stand sixth in significance.

Crickets were the response. Nothing. No news but insects outside the room.

How could a very junior member of the people in the room take control? It was a room in which five or six, if you included me, would ultimately win Nobel Prizes? With support from the Bills, I slept well that night.

It was obvious to me, though a complete surprise when I presented Table 3.1, that there were two serious scientists on the Committee who had profound interests in the table. They disagreed on the mean and median estimates of climate sensitivity. They were both looking at the possibility that a poor showing in an Academy report would place government support for their competing research programs at risk. They were right to worry. They both lost their funding when our report was issued.

The committee went out to dinner in the middle of the airborne fraction debate to a restaurant located on the first floor of the Watergate complex. It was a wonderful dinner – we were staying at the nearby River Inn as is what was then the NAS. That is to say, our lodging was usual, comfortable and not extravagant. I would confirm this opinion without any complaint while working on many more Academy reports - but this experience was my first.

The dinner was spectacular. Chair Nierenberg showed off his knowledge of wine by ordering many offerings (one for each for many courses). He was enough of a big deal that the Academy paid for the entire meal, including the wine (the only time in my life that I have such a thing happen except when I served as host).

Perhaps most importantly, it was the first time that I had ever tasted cinnamon ice cream for dessert – still a favorite of mine, much to the delight of my children and now my grandchildren.

Authors contributing to the climate literature are still citing our chapter, in part because the summaries illustrated in our spaghetti graphs are still inside the norm.

For example, well before its Sixth Assessment Report and based in large measure on our Chapter, the Intergovernmental Panel on Climate Change made it clear that communicating projected (not predicted) changes in climate had moved beyond concentrations to projections of global mean temperature. That would ultimately link temperature change with Reasons for Concern (the five RFCs) – and so impacts that would move us beyond focusing on concentrations. More on that, later.

Temperatures are, of course, themselves linearly related to atmospheric concentrations, which are a direct function of cumulative emissions. Since our equations reflected those findings, our work, remained relevant for many years.

William Clark of Harvard later commented well into this century – "have we had not progressed at all?" He remembered well the 1982 piece to which he had contributed more than 15 pages of review comments – the most copious and forthright review I have ever seen.

Notwithstanding assessments' moving one step down the causal chain to temperature, Figure 3-2 shows that Figure 3-1 is still representative of at least some of what we know in 2023 about how humans drive climate change. Only now, in 2023, are emissions for three different greenhouse gases tracked and projected along at least four alternative visions of how the future socio-economic environments of the globe might evolve through to 2100. Now we are more worried more than ever before about tipping points that might trigger unforeseeable possibilities that had been buried within RFC5.

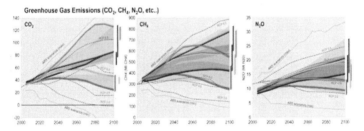

**Figure 3-2. Projected emissions ranges with medians highlighted in bold for major heat-trapping gases: carbon dioxide, methane, and nitrous oxides.**
Source: https://www.sciencedirect.com/science/article/pii/S0959378016300681

# References, abstracts and links connected to my January 2023 CV

**#10** Nordhaus, W. and Yohe, G., "Future Carbon Dioxide Emissions from Fossil Fuels," in *Changing Climate: Report of the Carbon Dioxide Assessment Committee*, National Research Council, Washington, 1983.
This section deals with the uncertainty about the buildup of C02 in the atmosphere. It attempts to provide a simple model of C02 emissions, identify the major uncertain variables or parameters influencing these emissions, and then estimate the best guess and inherent uncertainty about future C02 emissions and concentrations. Section 2.1.1 is a self-contained overview of the method, model, and results. Section 2.1.2 contains a detailed description of sources, methods, reservations, and results

**#12** Yohe, G., "The Effects of Changes in Expected Near-term Fossil Fuel Prices on Long-term Energy and Carbon Dioxide Projections" *Resources and Energy* **6**: 1-20, 1984.
Two surveys of near-term energy related projections produced thus far by the International Energy Workshop have produced two different views of the world through the year 2000. This paper considers the effects of these differences on long-term forecasts of global energy consumption and corresponding emissions of carbon dioxide. The lower growth forecast of the more recent survey portends lower economic activity in the next century, but carbon emissions of nearly the same dimension. Slower growth can, in particular, be expected to slow the price induced substitution into non-fossil fuels so that the income effect of slower growth is balanced by a slower, less vigorous substitution effect.

**#13** Yohe, G., "Constant elasticity of substitution production functions with three or more inputs: and approximation procedure", *Economics letters*, 1984.
An approximation procedure is developed to allow arbitrary and constant elasticities of substitution between aggregate inputs in production functions with three or more factors. The procedure is necessary to overcome the serious constraints on structure imposed on CES schedules by the Uzawa-McFadden results.

**#15** Yohe, G., "Evaluating the efficiency of long-term forecasts with limited information: Revisions in the IEW poll responses", *Resources and Energy* **8**:331-339, 1986.
An analytical technique based on an adaptive expectations model or incorporating current information into long-term forecast is developed and applied to International Energy Workshop respondents' revisions or reported oil price forecasts. Current year weights implicit in their revisions have mean values ranging from 0.27 through 0.67 depending upon the forecast horizons and the assumed base cases. Since the revisions suggest greater perceived importance or current changes for longer forecast, though, some doubt is cast upon the efficiency or their revisions.

**#19** Yohe, G., "Uncertainty, Global Climate and the Economic Value of Information", *Policy Sciences* 24: 245-269, 1991.
Increased atmospheric concentrations of radiatively active gases (e.g., carbon dioxide, various chlorofluorocarbons, methane, nitrous oxides, etc...) are expected to cause mean global temperatures to rise 2 °C to 5 °C over the next century. The effects of this greenhouse warming are likely to be widespread, but our current understanding of their potential dimension and their ultimate social, economic and political impacts is clouded with enormous uncertainty. In light of disagreement about how much and where, a recent Report to Congress by the U.S. EPA (1989) advanced a range of greenhouse induced sea level rise through the year 2100 running from 50 centimeters on the low side to 150 centimeters on the high side; note that the upper boundary of the EPA range is 200% larger than the lower boundary and is deemed equally likely on the basis of the best information available in 1989.

**#21** Yohe, G., "Selecting 'Interesting' Scenarios with which to Analyze Policy Response to Potential Climate Change" *Climate Research* 1: 169-177, 1991.
The interactive complexity of the sources and effects of global climate change make it nearly impossible to analyze adaptive and/or abatement response strategies across the full range of possible futures. Researchers are therefore left to investigate their options along a few "interesting" scenarios of what the future "might" hold, and so they must focus early attention upon how to select those scenarios Two selection metrics are examined here. One, based upon minimizing the subjective mean squared error of when some critical state variable (or vector of variables) might cross a response threshold. It has its roots in elementary statistical theory, but it is purely a physical measure. A second criterion, based upon minimizing the subjective expected cost of responding at such a threshold, has the advantage of incorporating these costs into the selection process. Subsequent analysis of the potential benefits of abatement along scenarios selected in application of the second metric can thereby incorporate a reflection of efficient adaptation - a first step towards seeing how to strike an appropriate balance between adaptation and abatement.

**#31** Yohe, G., and *Wallace, R. "Near Term Mitigation Policy for Global Change Under Uncertainty - Minimizing the Expected Cost of Meeting Unknown Concentration Thresholds", *Energy Modeling Assessment* 1: 47-57, 1996.
An aggregate integrated assessment model is used to investigate the relative merits of hedging over the near term against the chance that atmospheric concentrations of carbon dioxide will be limited as a matter of global policy. Hedging strategies are evaluated given near term uncertainty about the targeted level of limited concentrations and the trajectory of future carbon emissions. All uncertainty is resolved in the year 2020, and strategies that minimize the expected discounted value of the long term cost of abatement, including the extra cost of adjusting downstream to meet unexpected concentration limits along unanticipated emission trajectories, are identified. Even with uncertainties that span current wisdom on emission futures and restriction thresholds that run from 550 ppm through 850 ppm, the results offer support for at most modest abatement response over the next several decades to the threat of global change.

**#40** Yohe, G. and Schimmelpfennig, D., "Vulnerability of Agricultural Crops to Climate Change: A Practical Method of Indexing" in *Global Environmental Change and Agriculture: Assessing the Impacts*, Edward Elgar Publishing, 1999.
We begin with the notion that the probability of crossing a threshold can be a workable metric of vulnerability. The idea of action thresholds was proposed by participants in a landmark international conference held in Villach, Austria (SCOPE, 1985) and it has been emphasized again in the highly visible Intergov-ernmental Panel on Climate Change (IPCC) assessment reports (IPCC, 1996). This chapter will add to that discussion and to our knowledge about thresholds by developing a uniformly applicable index to characterize probabilistically, the crossing of one or more thresholds. The vulnerability index accounts for uncer-tainty in our understanding of how the climate might be changing and uncertainty in our

understanding of the consequences of climate change. A complementary index of sustainability is simply one minus the vulnerability index.

**#41** Yohe, G., *Jacobsen, M., and *Gapotochenko, T., "Spanning 'Not Implausible' Futures to Assess Relative Vulnerability to Climate Change and Climate Variability", *Global Environmental Change* **9**: 233-249, 1999.

This paper offers a framework of a simple method designed specifically to help the research community come to grips with this selection issue. It suggests a way that we might build on simple "first generational impact/adaptation analyses to determine how various sources of cascading uncertainty might alter our understanding of how to plan for what the future might hold. The idea is to apply vulnerability indexing schemes like the one developed recently by Schimmelpfennig and Yohe (1998) to some existing case studies of vulnerable systems (or even a heuristic description of sources of stress on systems thought to be vulnerable). Each study or description will have identified critical impact variables that drive the relevant climate impact and frame the associated adaptation questions.

It may have worked through some of the possible adaptive strategies for a few climate change scenarios. Perhaps it will have discarded some adaptive strategies that are not adoptable given the specifics of their systems' cultural, socio-economic and political structures. Perhaps it will have looked into the informational constraints that limit the potential efficacy of those strategies. Perhaps it will have placed climate stress and adaptation into the context of the systems' anticipated responses to other stresses. Perhaps it will have incorporated climate variability into their considerations. Even if none of these complications has been considered, the method described here requires only that existing studies or stories identify the critical impact variables that drive prospective change and the context of a possible adaptive response. Indeed, is to see if it would be more beneficial to expand our existing knowledge of how a given system might work as the future unfolds or simply to move on to consider another system, altogether.

**#52** Strzepek, K., Yates, D., Yohe, G., Tol, R. and *Mader, N., "Constructing "Not Implausible" Climate and Economic Scenarios for Egypt", *Integrated Assessment* **2**: 139-157, 2001.

A space of "not-implausible" scenarios for Egypt's future under climate change is defined along two dimensions. One dimension depicts representative climate change and climate variability scenarios that span the realm of possibility. Some would not be very threatening. Others portend dramatic reductions in average flows into Lake Nassar and associated increases in the likelihood of year to year shortfalls below critical coping thresholds; these would be extremely troublesome, especially if they were cast in the context of increased political instability across the entire Nile Basin. Still others depict futures along which relatively routine and relatively inexpensive adaptation might be anticipated.

# CHAPTER 4

# SEA LEVEL RISE

*The lede here is that sea level rise (SLR) was my first foray into impacts and adaptation. In a developed country like the United States, coastal locations are the perfect laboratory – detection reveals that rising seas and attribution to warming temperatures are solid. The baseline source of rising seas is the thermal expansion of the water that covers 75% of the earth's surface. However, local acceptance of either detection or attribution is all over the map. Many of my most cited papers are on sea level rise.*

My earliest work on sea level rise recognized the range of assumptions. On one hand of two extremes was an assumption consistent with the "dumb farmer" in agriculture: individuals who did not recognize any changes in their circumstances, and therefore did nothing to respond until it was too late to do anything but the most expensive option. The other extreme involved "smart markets" that would autonomously incorporate all available information so that both sides of the market had perfect information to enact the response that would be the least expensive option. That approach produced two cost estimates. I assumed that they were limiting estimates on the high and low sides, because actual decision-making behavior would lie somewhere in between.

Later work actually put humans and their institutions into the mix, with particular emphasis on New York City. Sea level rise and coastal zone management, especially given the potential for intense coastal storms (not only hurricanes), was still the perfect laboratory within which to include adaptation. Indeed, they were better reflections of what might be reality - autonomous and/or anticipatory adaptation (comparing, for example, Hurricanes Harvey, Irma and Maria in 2017 in terms of damage calibrated along many dimensions from currency to human lives estimates).

At the same time, my understanding evolved to include losses from storm surge - even for large but regular storms. Jason West and Hadi Dowlatabadi were among the first to put this on the table with a case study of the outer banks of North Carolina. As usual, Hadi was spot on.

I focused on economic metrics and C-B motivated adaptations for a long time, but I eventually began to worry about social policies and "tolerable risk" thresholds written and articulated by human decision-makers. These conventions define several levels of the determinants of adaptive capacity: the ability to separate signal from noise, the availability to response options and resources, the willingness to accept decision-making responsibility, and the credibility of social and political capital constructs.

There will be more on adaptive capacity in chapter 7, because it is in this context that it became clear that taking account of the "dark tails" a la Steve Schneider is critical for decision-makers and opinion-makers whose tasks include protecting populations around the world from climate risk. Descriptions of some of the tails had emerged from assessments of well-established, "not-implausible" possibilities. They were hereby alerted that managing risk would include contemplating extremes. Some responded accordingly. Others did not, and so they exposed themselves to what could be very expensive if not catastrophic.

In either case, people should not necessarily prepare to protect against the historical record's worst event; that would mean accepting a level of tolerable risk as close to zero as current data could characterize.

Society cannot afford to do that in the climate world where it has become clear that the past is NOT a prologue to the future. Nor should decision-makers simply prepare for protecting against the median expectation of the future. If they accept any non-zero possibility that social definitions of "tolerable risk" could be violated by something that might not-implausibly occur sometime in the future, then they must take that into account.

This was the welcome conclusion of Mayor Bloomberg of New York City, who had made billions of dollars managing financial risk and would eventually see climate as just another source of financial and human danger. He created the New York (City) Panel on Climate Change (NPCC) in 2010. Its first year was spent, in large measure, exploring how to think about climate change and its potential damages. It was still a work in progress around 2011 when the NPCC emissary to the Mayor (his Chief of Staff) took his time in an elevator ride from the basement of City Hall up to the Mayor's office to convince him that climate change was a risk management problem.

The Mayor agreed; he said "make it happen", and that broke a thought logjam in the five boroughs. At least he got it.

The world had started to come to the same conclusion in 2007 when the IPCC wrote, in the *Summary for Policymakers of the Synthesis Report of the Fourth Assessment* that "[r]esponding to climate change involves an iterative risk management approach including...." This is the topic of chapters 9 and 10.

Once that happened, things changed around the world. Once this happened in NYC, evacuation plans changed. Historically, the plan in the face of a large and imminent coastal storm was to get millions of New Yorkers to higher ground by subway. Immediately, the plan changed. They were now to go to the nearest higher ground – the third floor of the nearest tall building.

By executive order, as a result of this insight, the Mayor shut down the subways 8 hours before Hurricane Sandy made landfall on October 28th of 2012. In Manhattan, this meant that 10 trains with 10 cars each carrying 100 passengers were not in flooded tunnels at the height of the storm. At least ten thousand people were waiting things out on the third floors of the highest nearby building. He had saved at least 10,000 lives – not to mention hundreds of thousands of dollars in equipment and electronic damage that was avoided because the train cars did go to higher ground (empty) with plenty of warning and no passengers.

Here is an interesting fact from a different context: a quadratic SLR cost function had been widely accepted as the standard form in calibrating economic damages as a function of temperature change in many integrated assessment models. It was part of paper #47, but only as footnote, as it was never intended to set a more broadly defined standard. It was an estimate based on dumb or clairvoyant markets for a sample of developed property scattered along the coastline of the United States; it was never meant to be an estimate or even a form of a damage function that would apply more broadly.

For reference, Figures 4-1 through 4-5 show what SLR risk results looked like at the turn of the century in terms economic loss across coastal areas in the U.S. Figure 4-1 illustrates a standard modeling schematic from the 1990s. Figure 4-2 replicates results for carbon dioxide emissions reported in #37, derived by the procedures described in #21 and refined according to #31 to produce a limited collection of "interesting", not implausible, and representative scenarios, whose ranges reflect a wide distribution of outcomes from thousands of Monte-Carlo simulations.

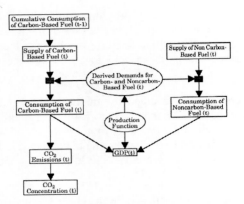

**Figure 4-1. A schematic of a global emissions model.** Source: Figure 1 in #37 below.

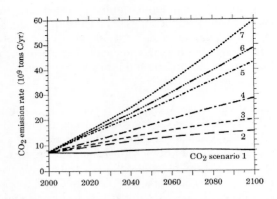

**Figure 4-2. Representative global carbon dioxide emissions scenarios.** Source: Figure 2 in #37 derived by the procedure described in #21 and expanded in #31.

Figure 4-3 translates those results into atmospheric concentrations (and thus global mean temperatures) before we get to the cost estimates. Those numbers are reflected in Figures 4-4 and 4-5. The first gives the probabilistic range for costs over time for a complete set of deciles. The second is likely to be more interesting, since it is an early representation of the value of conveying SLR information to the coastal real estate markets, so that market value can reflect risk. The gap between the lines represents the reduced cost from coastal vulnerability if the demand side of the markets fully and accurately incorporated the then-current understanding of the pace of rising oceans into their evaluation of exposed properties. This helps in making the decision of whether or not to buy personal property, and whether or not to continue to invest in maintaining it, given a future that includes the real probability of abandonment.

Bringing this analysis into the 21st century, the two panels of Figure 4-6 show five regional damage estimates around the coastline of the United States along four different mitigation futures through 2100. Superimposed on a map of the U.S. to identify the region, the graphs are calibrated in dollars of loss per dollar of coastal property. This makes for a very difficult comparison to earlier estimates that were not regional. Table 4-1 provides the information for that comparison by reporting U.S. aggregates for those regional estimates in 2020 dollars within four different climate eras identified by their middle years (for example, 2030 is the climate era between 2020 and 2040, and so on).

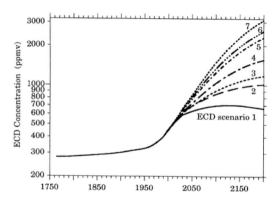

**Figure 4-3. Corresponding representative global carbon dioxide concentrations.**
Source: Figure 3 in #37.

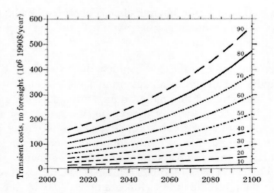

**Figure 4-4. Percentile specific expected costs from SLR over time and across the United States with no foresight in $millions (1990).** Source: Figure 7 in #37

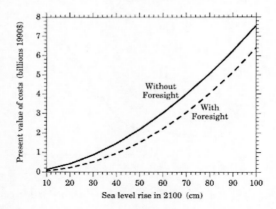

**Figure 4-5. Present value of expected US costs without and with foresight in $billions (1990) for different levels of SLR through 2100.** Source: Figure 8 in #37.

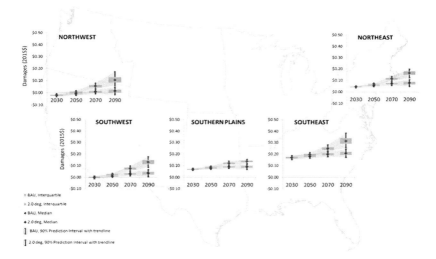

**Figure 4-6. Panel A Regional coastal losses along a business as usual scenario (the higher ranges) or a mitigation trajectory that holds increases in global mean temperature below 2 degrees Celsius.**

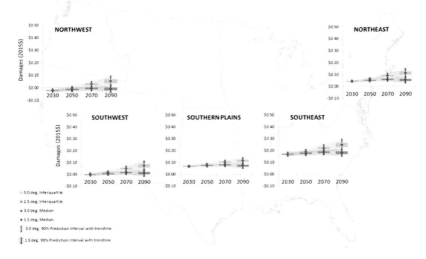

**Figure 4-6 Panel B. Regional coastal losses mitigation along trajectories that hold increases in global mean temperature below 1.5 (the lower ranges) or 3 degrees Celsius, respectively.**

**Table 4-1. Total annual damages to coastal property across the U.S. coastline along the contiguous 48 states ($1b).** Source: Table 2 in #37.

| Climate Era | 1.5 Degrees | 2.0 Degrees | 3.0 Degrees | BAU |
|---|---|---|---|---|
| 2030 | $4.6 | $4.7 | $4.9 | $5.0 |
| 2050 | $5.0 | $5.5 | $6.3 | $7.2 |
| 2070 | $5.3 | $6.3 | $8.3 | $12.0 |
| 2090 | $5.0 | $6.3 | $9.4 | $20.0 |

While these estimates are comparable, they are sufficiently different to warrant an explanation. The key is understanding that these estimates are calibrated in different dollars – 1990 versus 2020. The CPI nearly doubled over those decades; real estate prices and energy costs lead that rise, and coastal property became enormously attractive (read high demand and high prices). Perhaps most importantly, the regional estimates from 2020 captured differences in historical and projected SLR across locations; the places where local SLR would be the highest also happened to be the places where the coastal investment in property and amenities has been the largest.

# References, abstracts and links connected to my January 2023 CV

**#18** Yohe, G., "The Cost of Not Holding Back the Sea – Economic Vulnerability", *Ocean Shoreline Management* 15: 233-255, 1989.
A method for quantifying the economic vulnerability of developed shoreline to the threat of greenhouse induced sea level rise is described and applied to Long Beach Island, New Jersey, USA. While the method carefully accounts for structure, land and beach vulnerability along arbitrary sea level rise scenarios from tax maps and careful geographical accounting, it does not produce opportunity cost estimates for abandonment. The data generated here are, nonetheless, the foundation from which such cost estimates can be constructed given market and individual reactions to subjective perceptions of the threat and its timing.

**#20** Yohe, G., "The Cost of Not Holding Back the Sea – Towards a National Sample of Economic Vulnerability", *Coastal Management* 18: 403-431, 1991.
https://www.tandfonline.com/doi/abs/10.1080/08920759009362123
National and regional estimates of U.S. economic vulnerability to greenhouse-induced sea-level rise are produced from a sample of 30 discrete regions scattered evenly along the coastline. Scenarios that envision 50 cm, 100 cm, and 200 cm of greenhouse-induced sea-level rise are considered. They can be expected to place $39.2, $65.6, and $133.3 billion, respectively, (1989 dollars) of existing development in jeopardy through 2050, and $133.3, $308.7, and $909.4 billion through 2100.

Sampling error and consideration of the uncertainty with which we currently view future greenhouse-induced sea-level rise places the 25th and 75th percentile values of expected cumulative vulnerability at $38.5 and $76.7 billion through 2050 and $132.6 and $362.4 billion through 2100. Not surprisingly, the southeast displays the largest potential vulnerability, with the northeast ranking second above both the Gulf coast and the west coast.

**#22** Titus, J.G., Park, R.A., Leatherman, S.P., Weggel, J.R., Greene, M.S., Mausel, P.W., Brown, s., Gaunt, G., Threhan, M. and Yohe, G., "Greenhouse Effect and Sea Level Rise: the Cost of Holding Back the Sea", *Coastal Management* 19: 171-204, 1991.

Previous studies suggest that the expected global warming from the greenhouse effect could raise sea level 50 to 200 centimeters (2 to 7 feet) in the next century. This article presents the first nationwide assessment of the primary impacts of such a rise on the United States: (1) the cost of protecting ocean resort communities by pumping sand onto beaches and gradually raising barrier islands in place; (2) the cost of protecting developed areas along sheltered waters through the use of levees (dikes) and bulkheads; and (3) the loss of coastal wetlands and undeveloped lowlands. The total cost for a one-meter rise would be $270-475 billion, ignoring future development. We estimate that if no measures are taken to hold back the sea, a one meter rise in sea level would inundate 14,000 square miles, with wet and dry land each accounting for about half the loss. The 1500 square kilometers (600-700 square miles) of densely developed coastal lowlands could be protected for approximately one to two thousand dollars per year for a typical coastal lot. Given high coastal property values, holding back the sea would probably be cost-effective. The environmental consequences of doing so, however, may not be acceptable. Although the most common engineering solution for protecting the ocean coast pumping sand would allow us to keep our beaches, levees and bulkheads along sheltered waters would gradually eliminate most of the nation's wetland shorelines. To ensure the long-term survival of coastal wetlands, federal and state environmental agencies should begin to lay the groundwork for a gradual abandonment of coastal lowlands as sea level rises

**#29** Yohe, G., Neumann, J. and Ameden, H., "Assessing the Economic Cost of Greenhouse Induced Sea Level Rise: Methods and Applications in Support of a National Survey", *Journal of Environmental Economics and Management* 29: S-78-S-97, 1995.

Estimates of the true economic cost that might be attributed to greenhouse-induced sea- level rise on the developed coastline of the United States are offered for the range of trajectories that is now thought to be most likely. Along a 50-cm sea level rise trajectory (through 2100), for example, transient costs in 2065 (a year frequently anticipated for doubling of greenhouse-gas concentrations) are estimated to be roughly $70 million (undiscounted, but measured in constant 19905). More generally and carefully cast in the appropriate context of protection decisions for developed property, the results reported here are nearly an order of magnitude lower than estimates published prior to 1994. They are based upon a calculus that reflects rising values for coastal property as the future unfolds, but also includes the cost-reducing potential of natural, market-based adaptation in anticipation of the threat of rising seas and/or the efficiency of discrete decisions to protect or not to protect small tracts of property that will be made when necessary and on the (then current) basis of their individual economic merit.

**#30** Yohe, G., Neumann, J., Marshall, P., and Ameden, H., "The Economic Cost of Greenhouse Induced Sea Level Rise in the United States", *Climatic Change* 32: 387-410, 1996.

Estimates of the true economic cost that might be attributed to greenhouse-induced sea- level rise on the developed coastline of the United States are offered for the range of trajectories that is now thought to be most likely. Along a 50-cm sea level rise trajectory (through 2100), for example, transient costs in 2065 (a year frequently anticipated for doubling of greenhouse-gas concentrations) are estimated to be roughly $70 million (undiscounted, but measured in constant 19905). More generally and carefully cast in the appropriate context of protection decisions for developed property, the results reported here are nearly an order of magnitude lower than estimates published prior to 1994. They are based upon a calculus that reflects rising values for coastal property as the future unfolds, but also includes the cost-reducing potential of natural, market-based adaptation in anticipation of the threat of rising seas and/or the efficiency of discrete decisions to protect or not to protect small tracts of property that will be made when necessary and on the (then current) basis of their individual economic merit.

**#33** Yohe, G. and Neumann, J., "Planning for Sea-level Rise and Shore Protection under Climate Uncertainty", *Climatic Change* 37: 243-70, 1997.

Attention is focused here on the effect of additional sources of uncertainty derived from climate change on the cost-benefit procedures applied by coastal planners to evaluate shoreline protection projects, The largest effect would be felt if planners were trying to achieve the first best economic optimum. Given the current view that the seas will rise by significantly less than one meter through the year 2100, present procedures should work reasonably well assuming (I) informed vigilance in monitoring the pace of future greenhouse induced sea level rise, (2) careful attention to the time required for market-based adaptation to minimize the economic dislocation, (3) firm support of tile credibility of an announced policy to proceed with plans to retreat from the sea when warranted assumptions (1) and (2) might be satisfied in reality. In any case, planners should periodically revisit potential protection sites, especially in the wake of catastrophic events, to assess the impact of the most recent information on sea level rise trajectories, local development patterns, and protection costs on the decision calculus.

**#37** Yohe, G. and Schlesinger, M., "Sea Level Change: The Expected Economic Cost of Protection or Abandonment in the United States", *Climatic Change* 38: 447-472, 1998.

Three distinct models from earlier work are combined to: (1) produce probabilistically weighted scenarios of greenhouse-gas-induced sea-level rise; (2) support estimates of the expected discounted value of the cost of sea-level rise to the developed coastline of the United States, and(3) develop reduced-form estimates of the functional relationship between those costs to anticipated sea-level rise, the cost of protection, and the anticipated rate of property-value appreciation. Four alternative representations of future sulfate emissions, each tied consistently to the forces that drive the initial trajectories of the greenhouse gases, are considered. Sea-level rise has a nonlinear effect on expected cost in all cases, but the estimated sensitivity falls short of being quadratic. The mean estimate for the expected discounted cost across the United States is approximately $2 billion (with a3% real discount rate), but the range of uncertainty around that estimate is enormous; indeed, the 10thand 90th percentile estimates run from less than $0.2 billion up to more than $4.6 billion. In addition, the mean estimate is very sensitive to associated sulfate emissions; it is, specifically, diminished by nearly 25% when base-case sulfate emission trajectories are considered and by more than 55% when high-sulfate trajectories are allowed.

**#39** Yohe, G., Neumann, J. and Marshall, P., "The Economic Damage Induced by Sea Level Rise in the United States" in *The Impact of Climate Change on the United States Economy* (Mendelsohn, R. and Neumann, J., eds.), Cambridge: Cambridge University Press, 1999.

Changes in climate are expected to affect the ocean environment in a variety of ways. The potential effects of a temperature increase include thermal expansion and the melting of polar ice caps, both of which contribute to the causes of sea level rise. Increases in sea level can present problems to people living in coastal and low-lying areas, and can damage structures and beachfront property along the coast. Consequently, a sea level rise may impose economic costs on the United States – the costs of protecting coastal structures and the shoreline, or the lost value associated with abandoning such structures and property.

Early predictions of dramatic greenhouse gas-induced sea level rise have given way over the past decade to more modest expectations. High projections for the year 2100reached more than 3.5 meters as late as 1983 (Hoffman et al., 1983), but they dropped to 1.5 meters in 1990 (IPCC, 1990), and converged slightly more than 1 meter by 1992(IPCC, 1992). The mid-range best guess now stands between 38 and 55 cm by 2100(IPCC, 1996). One recent estimate is presented in Table 7.1 (Wigley, 1995; Wigley and Raper, 1992). The oceans would continue to rise for centuries, even if concentrations were stabilized in the interim. Despite this, the highest best guess reported for the year2100 is 40 cm. One important contribution of this chapter is to present economic cost estimates for these new lower trajectories (less than 1 meter).

**#47** Neumann, J., Yohe, G., Nicholls, R., Manion, M., "Sea-Level Rise & Global Climate Change: A Review of Impacts to U.S. Coasts", Washington, D.C.: Pew Center on Global Climate Change, 2000.

This report finds that the vulnerability of a coastal area to sea-level rise varies according to the physical characteristics of the coastline, the population size and amount of development, and the responsiveness of land-use and infrastructure planning at the local level. The authors conclude the following:

> Low-lying developed areas in the Gulf Coast, the South, and the mid-Atlantic regions are especially at risk from sea-level rise.

> The rapid growth of coastal areas in the last few decades has resulted in larger populations and more valuable coastal property being at risk from sea-level rise. This growth, which is expected to continue, brings with it a greater likelihood of increased property damage in coastal areas.

> The major physical impacts of a rise in sea level include erosion of beaches, inundation of deltas as well as flooding and loss of many marshes and wetlands. Increased salinity will likely become a problem in coastal aquifers and estuarine systems as a result of saltwater intrusion.

> Although there is some uncertainty about the effect of climate change on storms and hurricanes, increases in the intensity or frequency or changes in the paths of these storms could increase storm damage in coastal areas.

Damage to and loss of coastal areas would jeopardize the economic and ecological amenities provided by coastal wetlands and marshes, including flood control, critical ecological habitat, and water purification. Damages and economic losses could be reduced if local decision-makers understand the potential impacts of sea-level rise and use this information for planning.

The authors and the Pew Center gratefully acknowledge Dr. Donald Boesch for his contributions to this work and our understanding of coastal risks from climate change. We mourn his untimely death (2000) swimming in Hawaii.

**#123** Yohe, G. and Leichenko, R., "Adopting a Risk-Based Approach", in New York City Panel on Climate Change, 2009, *Climate Change Adaptation in New York City: Building a Risk Management Response.* (C. Rosenzweig & W. Solecki, eds.), pp 29-40. Prepared for use by the New York City Climate Change Adaptation Task Force, *Annals of the New York Academy of Sciences,* New York, NY. 2009, p. 349. http://onlinelibrary.wiley.com/doi/10.1111/j.1749-6632.2009.05310.x/full.

A significant message accompanying the call for greenhouse gas mitigation actions from the Intergovernmental Panel on Climate Change (IPCC) 2007 Fourth Assessment Report is the increasing need to identify a decision framework for climate change that encompasses both mitigation and adaptation. Through the IPCC, governments have begun to acknowledge risk management as a unifying theme for both climate change mitigation and adaptation. Their unanimous approval of this message underscores the importance of providing more information about climate risks (in addition to providing information about impacts and associated vulnerabilities) and suggests that consideration of risk plays a critical role in all facets of climate change decision making:

> "Responding to climate change involves an ***iterative risk management process that includes both adaptation and mitigation***, and takes into account climate change damages, co-benefits, sustainability, equity and attitudes to risk" (IPCC, 2007c; our emphasis).

These words make clear that governments throughout the world now understand that managing the risks associated with climate change will be a central theme for present and future planning and policy decisions. For climate change adaptation, particularly in a large city like New York, a risk-based approach can serve as a valuable guide to policy and action. A critical aspect is that it can promote support of expenditure of evermore scarce city resources to reduce risks from both high-probability events and low-probability events.

**#140** Yohe, G., Knee, K. and Kirshen, P., "On the Economics of Coastal Adaptation Solutions in an Uncertain World", *Climatic Change* 106: 71-92, 2011.
The economics of adaptation to climate change relies heavily on comparisons of the benefits and costs of adaptation options that can range from changes in policy to implementing specific projects. Since these benefits are derived from damages avoided by any such adaptation, they are critically dependent on the specification of a baseline. The current exercise paper reinforces this point in an environment that superimposes stochastic coastal storm events on two alternative sea level rise scenarios from two different baselines: one assumes perfect economic efficiency of the sort that could be supported by the availability of actuarially fair insurance and a second in which fundamental market imperfections significantly impair society's ability to spread risk. We show that the value of adaptation can be expressed in terms of differences in expected outcomes damages only if the effected community has access to efficient risk-spreading mechanisms or reflects risk neutrality in its decision-making structure. Otherwise, the appropriate metric for measuring the benefits of adaptation must be derived from certainty equivalents. In these cases, increases in decision-makers' aversion to risk increase the economic value of adaptations that reduce expected damages and diminish the variance of their inter-annual variability. For engineering and other adaptations that involve significant up-front expense followed by ongoing operational cost, increases indecision-makers' aversion increase the value of adaptation and therefore move the date of economically efficient implementation closer to the present.

**141** Rosenzweig, C., Solecki, W., Gornitz, V., Horton, R., Major, D., Yohe, G., Zimmerman, R., "Developing Coastal Adaptation to Climate Change in the New York City Infrastructure-shed: Process, Approach, Tools, and Strategies", *Climatic Change* 106: 93-127, 2011.
While current rates of sea level rise and associated coastal flooding in the New York City region appear to be manageable by stakeholders responsible for communications, energy, transportation, and water infrastructure, projections for sea level rise and

associated flooding in the future, especially those associated with rapid ice melt of the Greenland and West Antarctic Ice Sheets, may be outside the range of current capacity because extreme events might cause flooding beyond today's planning and preparedness regimes. This paper describes the comprehensive process, approach, and tools for adaptation developed by the New York City Panel on Climate Change (NPCC) in conjunction with the region's stakeholders who manage its critical infrastructure, much of which lies near the coast. It presents the adaptation framework and the sea-level rise and storm projections related to coastal risks developed through the stakeholder process. Climate change adaptation planning in New York City is characterized by a multi-jurisdictional stakeholder–scientist process, state-of-the-art scientific projections and mapping, and development of adaptation strategies based on a risk-management approach.

**#213** Yohe, G., Willwerth, J., Neumann, J., and Kerrich, Z., 2020, "What the future might hold: Distributions of regional sectoral damages for the United States – Estimates and maps in an exhibition", *Climate Change Economics*, 11: 4, https://www.worldscientific.com/doi/10.1142/S2010007820400023
The text and associated Supplemental Materials contribute internally consistent and therefore entirely comparable regional, temporal, and sectoral risk profiles to a growing literature on regional economic vulnerability to climate change. A large collection of maps populated with graphs of *Monte-Carlo* simulation results support a communication device in this regard — a convenient visual that we hope will make comparative results tractable and credible and resource allocation decisions more transparent. Since responding to climate change is a risk-management problem, it is important to note that these results address both sides of the risk calculation. They characterize likelihood distributions along four alternative emissions futures (thereby reflecting the mitigation side context); and they characterize consequences along these transient trajectories (which can thereby inform planning for the iterative adaptation side).

Looking across the abundance of sectors that are potentially vulnerable to some of the manifestations of climate change, the maps therefore hold the potential of providing comparative information about the magnitude, timing, and regional location of relative risks. This is exactly the information that planners who work to protect property and public welfare by allocating scarce resources across competing venues need to have at their disposal — information about relative vulnerabilities across time and space and contingent on future emissions and future mitigation. It is also the type of information that integrated assessment researchers need to calibrate and update their modeling efforts — scholars who are exemplified by Professor Nordhaus who created and exercised the Dynamic Integrated Climate-Economy and Regional Integrated Climate-Economy models.

# CHAPTER 5

# MORE ON SEA LEVEL RISE –
# ADDING THE PROSPECT OF A TIPPING POINT
# IN ANTARCTICA

As shown in chapter 4, it is not difficult to demonstrate why global sea level rise (SLR) is a possible common source of enormous concern for any nation with developed coastlines distribution. Figure 5-1 displays the scientific basis supporting this assertion by displaying two distributions of future global sea level rise driven by anthropogenic emissions of carbon dioxide under two driving socio-economic cum mitigation contingencies.

The lower cones reflect the distribution of trajectories contingent upon an aggressive mitigation policy regime, with 5th and 95th percentile ranges over time surrounding a median trajectory that is depicted by a darker variegated line. The upper cones do the same for a high fossil fuel alternative scenario. For both cones, the difference between the light and dark regions shows the difference in projected ranges of possible SLR futures without (dark) and with (light) the recently detected extra contribution of melt-water from the West Antarctic Ice Sheet (WAIS).

For purposes of clarity, in what follows, we denote the trajectories which reflect recent data that show a doubling in the pace of the underlying trend since the turn of the 21st century, by "SLR+".

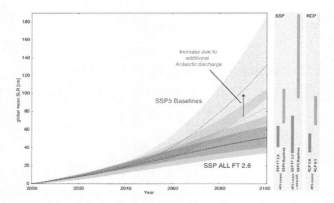

**Figure 5-1. Projections of global sea level rise from anthropogenic climatic change**. The distributions that populate the lower cones show the range of SLR projections through 2100 along a low emission scenario, supported by significant mitigation (Representative Concentration Pathway, RCP) 2.6 pathway, along which net emissions peak in the mid-2020s and turn negative in the second half of the century along five very distinct socio-economic pathways (SSPs).

http://www.gci.org.uk/Shared_Socio_Economic_Pathway.html. The two upper and lighter cones do the same for the high fossil fuel SSP5 baseline, which envisions rapid and unconstrained growth in economic output and energy use. Source: Alexander Nauels *et al* 2017 *Environ. Res. Lett.* 12 114002 and https://iopscience.iop.org/article/10.1088/1748-9326/aa92b6

Panel A of Figure 5-2 references the five distinct Shared Socioeconomic Pathways (SSPs) that were developed by extensive collaboration under the International Committee on New Integrated Climate Change Assessment Scenarios (ICONICS).[7] Panel B shows how the pathways span the space of possibilities with regard to the relative ease or difficulty with which various combinations of actions designed to combat climate risk might be designed, accepted, and eventually implemented. Some socio-economic characteristics favor one strategy; others favor another. For example, the sources of economic efficiency that favor mitigation can breed inequalities that challenge adaptation (SSP4). Conversely, a fossil fuel driven future that challenges mitigation with more difficult targets could also make adaptation responses more attractive (SSP5). SSP1, aptly dubbed "sustainability", favors both strategies. Another (SSP3), also aptly dubbed "regional rivalry", favors neither.

---

[7] See https://depts.washington.edu/iconics/

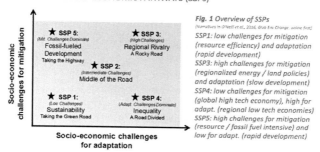

**Figure 5-2 Panel A. SSPs mapped into the challenges to space of contrasting challenges to mitigation or adaptation.** Figure 1 in
https://unfccc.int/sites/default/files/part1_iiasa_rogelj_ssp_poster.pdf

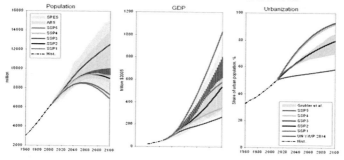

**Figure 5-2 Panel B. Ranges of population, GDP, and urbanization characterizing the SSPs.** Source: Replicating Figure 3 from
https://unfccc.int/sites/default/files/part1_iiasa_rogelj_ssp_poster.pdf

Finally, SSP2 captures a little of everything and so it suggests the summary lesson of the entire matrix: "abate, adapt, or suffer" are the only three options when it comes to climate risk. It follows that any choice is matter of degree. Mitigation reduces the likelihood of climate impacts, and adaptation reduces the consequences of the plethora of impacts that will only be delayed, and never completely avoided, by any degree of mitigation.

### The state of current tipping point climate science
### in the Antarctic (as of September of 2022)

The last ten years of research on the WAIS have shown an ominous development - a growing possibility that the mile-high columnar ice structures that have held the Thwaites glacier in place for many centuries may be disintegrating. If that new trend continues, it could suddenly and swiftly dump (in ten, or perhaps 15, years) enough water in the ocean to produce a foot or so of additional SLR above and beyond and SLR+ trajectory, in a matter of years rather than centuries. This possibility is reflected in Figure 5-2 by a single solid trajectory that tracks above the dotted upper median, moving increasingly father from it as the future unfolds; it is identified there by an arrow labeled "increase due to additional Antarctic discharge".

When might that happen? How fast will it happen after it has been triggered by an as yet undefined tipping point? How many other glacier locations could be expected to experience the same fundamental transition? Do we even know whether or not we have already committed the planet to the collapse, so that the question is really "when?" and not "if?". Are we sure that the glacier will not simply be stopped in its tracks for decades or centuries by ridges and other structures that populate the granite-glacier boundary that lies miles below the surface of the ice?

Then, there is the "so what?" question. We know that, depending upon local topography, coastal communities around the world are already being threatened by predictably rising oceans, now known to include larger than previously anticipated contributions from Antarctica (and Greenland). That is the point of Figure 5-1 – even with significant glacial melting in those locations, contingent future global SLR+ trajectories are sufficiently projectable to credibly inform adaptation decisions on the ground, and mitigation decisions at all relevant geopolitical, social and economic levels. The positioning of the projection contingencies on the matrix of Panel A of Figure 5-2 will suggest the degree to which those responses might work within various futures, and their broad social contexts can be inferred from Panel B.

Glacial vulnerability to collapse is a wild card born of an irreversible tipping point, the crossing of which could produce heretofore unimaginable additional SLR the world over in a matter of a decade or two; but how much confidence can be claimed in projections of its occurrence? Figure 3, a thought template created by the Intergovernmental Panel on Climate

Change (IPCC) to support authors' judgements in the fifth assessment (IPCC, 2014), offers a simple answer: not much when compared with our confidence in the trend distributions displayed in Panel A of Figure 2.

The template instructs us to consider agreement over scientific understanding of causality, as well as the quantity and quality of the evidence when judging confidence in a specific proposed finding. The divergence in SLR projections without and with additional contributions from Antarctic ice sheets is informed by increasingly robust data and high agreement with regard to attribution; "medium" is an appropriate assessment in this case.

By way of contrast, the potential rapid collapse of the Thwaites is informed by very limited data and medium agreement over the process; little more than "low" or even "very low" confidence can be claimed. By itself, though, this is not a reason to ignore this tipping point. As reported above, IPCC (2007) famously reported a synthetic finding that has framed nearly every evaluation of potential climate action since: "[r]esponding to climate change involves an iterative risk management process that includes both adaptation and mitigation" (page 22). The adjective "iterative" applied to "management" pointed to the expectation that mid-course corrections will be required. Inserting "risk" as a second modifier put the likelihood of a climate impact and its consequences front of mind. In that context, the complementarity relationship between mitigation and adaptation is clear: mitigation can reduce

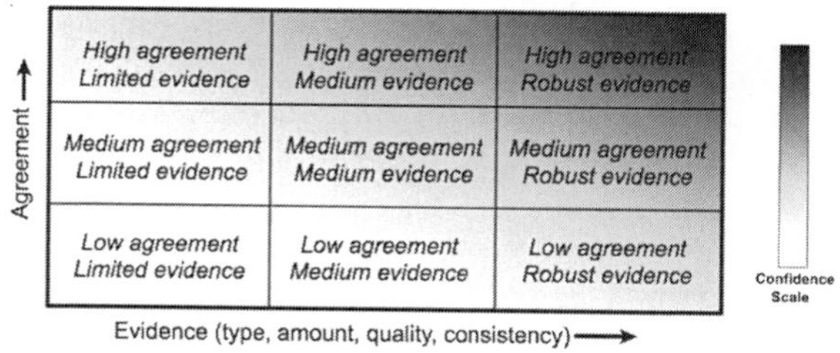

| High agreement<br>Limited evidence | High agreement<br>Medium evidence | High agreement<br>Robust evidence |
|---|---|---|
| Medium agreement<br>Limited evidence | Medium agreement<br>Medium evidence | Medium agreement<br>Robust evidence |
| Low agreement<br>Limited evidence | Low agreement<br>Medium evidence | Low agreement<br>Robust evidence |

Agreement →

Evidence (type, amount, quality, consistency) ⟶

Confidence Scale

**Figure 5-3. The joint influences of evidence and agreement on assessed confidence.** Increased confidence moving toward the upper-right corner is depicted by darker shadowing, very high confidence requires strength in both dimensions. Source: IPCC (2010)[8]

---

[8] https://www.ipcc.ch/site/assets/uploads/2017/08/AR5_Uncertainty_Guidance_Note.pdf

the likelihood of an impact, while adaptation can minimize the consequences. Finally, a society interested in abating high risk possibilities was directed to pay close attention to a high consequence event (like an ice sheet collapse), even if a very low likelihood or confidence level were attached.

## Depicting and reporting the Thwaites tipping point

The solid trajectory drawn above the dashed median pathway for the upper uncertainty cone for SLR+ in Figure 5-1 attracts attention to the existence of a tipping point for Thwaites, but it does not directly convey any of the necessary information for either the decision makers who must try to cope with the enlarged threat, or the investors who want to include all sources of risk into their own decisions. The line is not clearly defined. In fact, it cannot be described rigorously because it is neither a median nor a mean; the necessary data simply do not exist to support the either characterization. Moreover, it is not surrounded by its own uncertainty cone, and so it does not specifically convey the true character of a tipping point - an uncertain source of extra harm that produces sudden jagged departures from trend SLR projections of unknown size at an unknown future time.

Figure 5-4 displays this environment for three (of many) illustrative cases of not-implausible, sudden and sloped departures, relative to a dashed upper cone median trajectory (denoted SLR+) that has been included to place these possible futures in relation to the median trajectory shown inside the red cone of uncertainty in Figure 5-1. It also lies above a lower solid trajectory that similarly depicts the median for the blue uncertainty come (denoted SLR).

Three representative futures are displayed because timing is unknown across long periods of time, but is potentially important, depending on the consequences. The first jagged pathway depicts a case where the ice sheet collapsing tipping point lies 10 years into the future from the present. The next shows a tipping point in 30 years; and the rightmost, in 50 years. Other dates are surely possible, and the likelihood that one specific case will actually occur is, currently, very small. However, the likelihood that a crossing will have occurred before one specific temporal benchmark is potentially much larger and growing as the time horizon expands. Each pathway shows a sudden acceleration of the pace of SLR that lasts until the collapse has been completed; when that happens, the historical or projectably amplified global SLR trend can perhaps return (except that Thwaites may not be the only glacier in jeopardy of collapse - but that is another story). Besides the timing of such events, their durations (the horizontal dimension of the straight lines) and their associated cumulative extra SLR from the collapse, their vertical dimensions are also critical factors in driving their global significance.

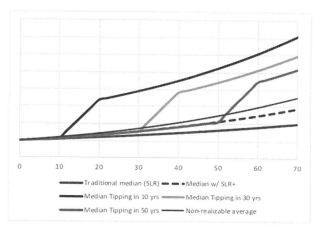

**Figure 5-4. Illustrative and representative SLR trajectories.** Illustrative SLR trajectories (versus years since 2023 absent uncertainty cones) for a traditional median with historical data (SLR), a median with extra contributions from Antarctic (SLR+), three tipping point possibilities built up from SLR+, and an average across all possible tipping point timing and significance.

Figure 5-4 can now support a discussion of the requirements of effective communication of tipping point risk from potential ice sheet collapse to companies and investors so that they can project their material risk in their annual risk reporting, but only by convincing decision makers, opinion makers, and investors that they have to look elsewhere for the information that they need. They have to look downstream in the causal chain to produce distributions of coastal damage contingent on prospective levels of SLR. Moreover, they need to look upstream to understand what is going on with the glacier on the ground. They then need to judge the value of that information at the downstream point of departure on the material risk that threatens upstream locations. The value of the upper cone of uncertainty in Figure 5-1 is simply showing that making this necessary connection is possible.

Specific informational needs can now be discussed. At a minimum, they include:

*Estimates of the extra global SLR contribution potential of this glacier or that (so that they can ultimately understand how big the problem will ultimately be at their specific coastal location);*

> Distributions of these estimates already exist for some locations like Thwaites, but their coverage should be expanded even while the underlying science improves agreement about underlying driving processes.

*As much understanding as possible of how long it would take the collapsing scenario to run its course from beginning (when its tipping point had been crossed) to end (when the full potential SLR had been realized);*

> Improved agreement about the processes that drive the glacier toward collapse can contribute here in at least two ways – clarifying the robustness of these critical duration estimates and strengthening the identification and measurement of the underlying drivers and potential impediments that regulate the collapse. The evolution of this understanding must, of course, be recorded and reported routinely without prejudice.

*Distributions of estimates of when the collapse might begin;*

> Quantifying these distributions is aspirational at this point, but a productive first step toward a "workaround" would involve characterizing how an irreversible collapse might be triggered, and what moves the glacier closer to those triggers. Progress there would identify the causal factors that could signal a glacier's approaching such a trigger point, and these factors could then be monitored physically on the ground to inform the evolution of downstream actions based on an iterative risk management approach.

*Socially accepted thresholds of tolerable risk at specific coastal locations, and timetable estimates for feasible adaptation plans designed to hold actual residual risk below those thresholds as the future unfolds;*

> All of this can be the product of contingency-based work that happens upstream and could produce estimates of

how long it would take to bring those adaptation plans to fruition, and whether or not their evolution, as SLR proceeds, could keep up with the pace of SLR+ as well as of the extra SLR contributions from glacial collapses.

Information like this could support comparisons of estimates of the total duration of the ice sheet collapse with estimates of the time it would take to bring an adaptation program from start to finish. As well as this, those comparisons could inform when various locations need to start adaptation projects contingent on actual observations from programs that monitor in situ the drivers of glacial movement toward tipping point triggers.

## Is there a potential for financial contagion?

Accurately informed markets are among the best protections from contagion that can otherwise cause macro scale economic harm – potential hazards can be revealed not only by exploring ranges of projectable trends, but also (and perhaps more importantly) by coming to grips with sudden and irreversible tipping points. To be convinced, simply recall that the US mortgage crisis of 2008, that brought about the great recession of 2008-9 and affected economic stability around the world, was fueled by opaque bundles of real estate derivatives whose values suddenly collapsed from artificial and unsustainable highs to near-zero lows. Forewarned by that experience, the Fed wrote in Box 4 of its Nov 9, 2020 "Financial Stability Report" that (Fed, 2020):

"Features of climate change can also increase financial system vulnerabilities.... Opacity of exposures and heterogeneous beliefs of market participants about exposures to climate risks can lead to mispricing of assets and the risk of downward price shocks. Similarly, uncertainty about the timing and intensity of severe weather events and disasters, as well as the poorly understood relationships between these events and economic outcomes, could lead to abrupt repricing of assets. Climate risks thus create new vulnerabilities associated with non-financial and financial leverage. In regions affected by severe events, households and businesses could become over-levered if the value of their assets or income prospects become impaired."

Later in the same box, the Fed looked at the opposite extreme when it wrote:

"With perfect information, the price of real-estate-linked assets and the valuations of claims linked to such assets—held by banks, insurers, investment funds, and nonfinancial firms—would reflect these climate-related risks".

Finally, the Fed made it clear that they had a specific climate-based contagion in mind – one derived from distinct global distributions of temporal scenarios of global sea level rise through 2100. Each element in the distribution would be produced by driving trajectories of anthropogenic emissions of greenhouse gases (GHGs), like carbon dioxide, that are contingent on specific socio-economic scenarios through models that reflect diverse challenges to both mitigation and adaptation responses. The news from that exercise is very troubling.

# CHAPTER 6

# REASONS FOR CONCERN

*Reasons for Concern (RFC,) and their illustrative "burning embers" diagram, were invented leading up the publication of the Third Assessment Report (TAR) of the Intergovernmental Panel on Climate Change (IPCC) in 2001. Their emergence from chapter 19 of the Report of Working Group II began a process of calibrating impacts and vulnerabilities in a variety of metrics: currency, species, lives in jeopardy or lost, and so on. The embers thereby laid the foundation for changing the analytic and assessment landscapes for scholars and, by implication, an increasing number of decision-making platforms around the world.*

The history of the IPCC's Reasons for Concern (RFCs) is perhaps most effectively traced by tracking the series of "burning ember" representations of the content of the underlying assessments that began in 2001 with the Third Assessment Report (TAR). Panel A of Figure 6-1 replicates its black and white representation in chapter 19 of the contribution of Working Group II. Subsequent, more colorful manifestations from the Synthesis Report of the TAR and beyond exist, but the black and white version is displayed in Panel B, along with temperature trajectories from then-current socio-economic scenarios.

Only two of the original "Reasons" targeted economic distributions and aggregate economic values measured in currency: risks associated with the distribution of impacts (RFC3) and risks associated with global aggregate impacts (RFC4, using current names from #171). That was big news back then because the others were drawn from different literatures: risks to unique and threatened systems (RFC1), risks associated with extreme weather events (RFC2), and risks associated with large-scale singular events (RFC5). It was such big news that insights drawn from this qualitative and subjective expansion of broad potential vulnerabilities were elevated to the Technical Summary and further to the Synthesis Report of the entire Fourth Assessment. It was new science, but the authors got away with it because the embers were so communicative.

The latest (and not the last, I am sure) iteration is shown in Panel C. Unlike the more obscure subjective authors' judgements that framed the first versions, the latest version takes explicit recognition of eight key risks (taken directly from Table 6-1 in O'Neill et al (#171)):

    i. Risk of death, injury, ill-health, or disrupted livelihoods in low-lying coastal zones and small island developing states and other small islands due to storm surges, coastal flooding, and sea-level rise;

    ii. Risk of severe ill-health and disrupted livelihoods for large urban populations due to inland flooding in some regions;

    iii. Systemic risks due to extreme weather events leading to breakdown of infrastructure networks and critical services such as electricity, water supply, and health and emergency services;

    iv. Risk of mortality and morbidity during periods of extreme heat, particularly for vulnerable urban populations and those working outdoors in urban or rural areas;

    v. Risk of food insecurity and the breakdown of food systems linked to warming, drought, flooding, and precipitation variability and extremes, particularly for poorer populations in urban and rural settings;

    vi. Risk of loss of rural livelihoods and income due to insufficient access to drinking and irrigation water and reduced agricultural productivity, particularly for farmers and pastoralists with minimal capital in semi-arid regions;

    vii. Risk of loss of marine and coastal ecosystems, biodiversity, and the ecosystem goods, functions, and services they provide for coastal livelihoods, especially for fishing communities in the tropics and the Arctic; and

    viii. Risk of loss of terrestrial and inland water ecosystems, biodiversity, and the ecosystem goods, functions, and services they provide for livelihoods.

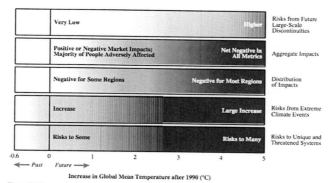

Figure 19-7: Impacts of or risks from climate change, by reason for concern. Each row corresponds to a reason for concern; shades correspond to severity of impact or risk. White means no or virtually neutral impact or risk, light gray means somewhat negative impacts or low risks, and dark gray means more negative impacts or higher risks. Global average temperatures in the 20th century increased by 0.6°C and led to some impacts. Impacts are plotted against increases in global mean temperature after 1990. This figure addresses only how impacts or risks change as thresholds of increase in global mean temperature are crossed, not how impacts or risks change at different rates of change in climate. Temperatures should be taken as approximate indications of impacts, not as absolute thresholds.

**Figure 6-1 Panel A. The first reasons for concern figure with its caption.** Source: Chapter 19 of the Report of Working Group II to the Fourth IPCC Assessment (reference #55).

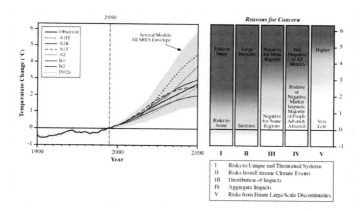

**Figure 6-1 Panel B. Later reasons for concern; imagine "burning embers" coloring in the right panel moving from white to yellow to bright red as warming runs up from 1 degree for RFCs I&II, up from 4 degrees for RFCs III & IV, and above 5 degrees for RFC V.** Or, go to the source: The right two panels are from Figure SPM-3 in https://www.ipcc.ch/site/assets/uploads/2018/05/SYR_TAR_full_report.pdf.

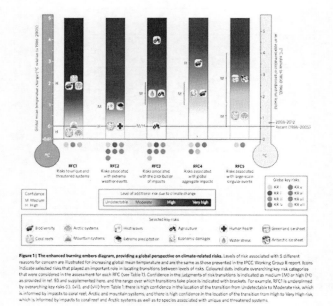

Figure 1 | The enhanced burning embers diagram, providing a global perspective on climate-related risks. Levels of risk associated with 5 different reasons for concern are illustrated for increasing global mean temperature and are the same as those presented in the IPCC Working Group II report. Icons indicate selected risks that played an important role in locating transitions between levels of risks. Coloured dots indicate overarching key risk categories that were considered in the assessment for each RFC (see Table 1). Confidence in the judgments of risk transitions is indicated as medium (M) or high (H) as provided in ref. 93 and supplemented here, and the range over which transitions take place is indicated with brackets. For example, RFC1 is underlined by overarching key risks (i), (vii), and (viii) from Table 1; there is high confidence in the location of the transition from Undetectable to Moderate risk, which is informed by impacts to coral reef, Arctic and mountain systems, and there is high confidence in the location of the transition from High to Very High risk, which is informed by impacts to coral reef and Arctic systems as well as to species associated with unique and threatened systems.

**Figure 6-1 Panel C. Reasons for concern with full captioning reported from the background documentation from the Fifth IPCC Assessment Report; purple (really dark black) was added for extreme concern.** Source: Figure 1 from #171.

The shading of each ember still provides a qualitative indication of the increase in risk, with higher temperatures for each individual "reason." Undetectable risk (nearly white) indicates that no associated impacts are detectable and attributable to climate change. Moderate risk (light gray) indicates that associated impacts are both detectable and attributable to climate change with at least medium confidence, also accounting for the other specific criteria for key risks. High risk (darker gray) indicates severe and widespread impacts, also accounting for the other specific criteria for key risks. Really dark gray almost black was introduced in the most recent assessment. It indicates very high risk, for which it was possible to refer to all eight specific criteria for all of key risks.

As noted in #126 and elaborated #175, reasons for concern have evolved over time. Each category was expanded in the Fourth Assessment Report of the IPCC (#87). I introduced a sixth "Reason" for the United States in #126: national security (FRC6). For example, among the original five:

- concern about risks to unique and threatened systems is no longer derived exclusively from natural systems; communities and other human systems that were threatened by climate change are now included in RFC1; and
- Distributions of impacts in RFC3 and RFC4 are now calibrated in metrics other than currency as long as they can be aggregated across nations (e.g., human lives at risk).

These changes began in the AR4 because of its emphasis across the contributions of Working Groups II and III to support risk management approaches to adaptation and mitigation. The assessments of the RFCs were supported by parallel application of a preliminary and anticipatory list of "key vulnerabilities" (magnitude, timing, persistence/irreversibility, the potential for adaptation, distributional aspects, likelihood, and importance) in chapter 19. The Synthesis Report of the entire Fourth Assessment Report (#85) again highlighted RFCs in the text, (pages 18-19), but the illuminating visual did not appear because of the strenuous objection of a major and influential country in North America. The image did appear, though, in #117 (Figure 1).

The Fifth Assessment Report (AR5) further advanced the application of RFCs (chapter 18), with better support on detection and attribution (#166) as well as increased global coverage in the impacts, adaptation and vulnerability literature (#167). These factors led to the list of "key risks" recorded above, but they also illustrated the sensitivity of the RFCs to emissions scenarios' futures. The time dimension of emissions was, though, still missing.

In #177, I worked to fill that gap by displaying as much of the #171 information as possible along distributions of transient temperature trajectories that I had created. Those distributions were tied to achieving 4 different temperature targets (along a median trajectory) in comparison with a no-policy baseline. It was thereby possible to infer the degree to which pursuing increasingly ambitious temperature targets could delay crossing thresholds of concern. Figure 6-2 shows representative results for RFC1 (risks to unique and threatened systems).

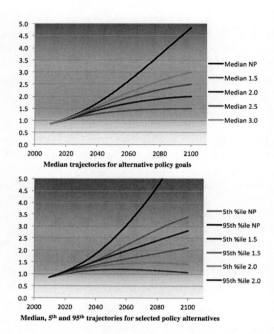

**Figure 6-2. Transient temperature scenarios displayed over the #171 judgements about levels of concern about RFC1** (risks to unique and threatened systems). Source: See Figure 4 in # 171 for the color version.

The conclusions that can be drawn from the median trajectories were expected, but the nuances of looking at distributions are noteworthy. The upper tails of transient temperatures from aggressive mitigation can be just as concerning as the medians of more lenient approaches. As well as this, there is a discernable difference between 1.5- and 2.0-degree targets – at least with regard to the timing of crossing thresholds of concern. Still, the 5th percentile shows the possibility that levels of concern might peak and then decline.

# References, abstracts and links connected to my January 2023 CV

**#55** Smith, J., Schellnhuber, J., Mirza, M., Fankhauser, S., Leemans, R., Erda, L., Ogallo, L., Pittock, B., Richels, R., Rosenzweig, C., Safriel, U., Tol, R.S.J., Weyant, J., Yohe, G., "Vulnerability to Climate Change and Reasons for Concern: A Synthesis", in *Climate Change 2001: Impacts, Adaptation and Vulnerability*, Cambridge: Cambridge University Press, 2001 (Section 19.7).

It does not appear to be possible — or perhaps even appropriate — to combine the different reasons for concern into a unified reason for concern that has meaning and is credible. However, we can review the relationship between impacts and temperature over the 21st century for each reason for concern and draw some preliminary conclusions about what change may be dangerous for each reason for concern. Note that the following findings do not incorporate the costs of limiting climate change to these levels. Also note that there is substantial uncertainty regarding the temperatures mentioned below. These magnitudes of change in global mean temperature should be taken as an approximate indicator of when various categories of impacts might happen; they are not intended to define absolute thresholds.

**#87** Bernstein, L., Bosch, P., Canziani, O., Chen, Z., Christ, R., Davidson, O., Hare, W., Huq, S., Karoly, D., Kattsov, V., Kundzewicz, Z., Liu, J., Lohmann, U., Manning, M., Matsuno, T., Menne, B., Metz, B., Mirza, M., Nicholls, N., Nurse, L., Pachauri, R., Palutikof, J., Parry, M., Qin, D., Ravindranath, N., Reisinger, A., Ren, J., Riahi, K., Rosenzweig, C., Rusticucci, M., Schneider, S., Sokona, Y., Solomon, S., Stott, P., Stouffer, R., Sugiyama, T., Swart, R., Tirpak, D., Vogel, C., and Yohe, G., *Climate Change 2007: Synthesis Report (for the Fourth Assessment Report of the Intergovernmental Panel on Climate Change)*, Cambridge: Cambridge University Press, 2007.

The five 'reasons for concern' identified in the TAR are now assessed to be stronger with many risks identified with higher confidence. Some are projected to be larger or to occur at lower increases in temperature. This is due to (1) better understanding of the magnitude of impacts and risks associated with increases in global average temperature and GHG concentrations, including vulnerability to present-day climate variability, (2) more precise identification of the circumstances that make systems, sectors, groups and regions that are especially vulnerable and (3) growing evidence that the risk of very large impacts on multiple century time scales would continue to increase as long as GHG concentrations and temperature continue to increase. Understanding about the relationship between impacts (the basis for 'reasons for concern' in the TAR) and vulnerability (that includes the ability to adapt to impacts) has improved (pages 64-65)

**#117** Smith, J.B., Schneider, S. H., Oppenheimer, M., Yohe, G., Hare, W., Mastrandrea,, M.D., Patwardhan, A., Burton, I., Corfee-Morlot, J., Magadza, C.H.D., Füssel, H-M, Pittock, A.B., Rahman, A., Suarez, A., and van Ypersele, J-P, "Dangerous Climate Change: An Update of the IPCC Reasons for Concern", *Proceedings of the National Academy of Science* 106: 4133-4137, March 17, 2009. Article 2 of the United Nations Framework Convention on Climate Change commits signatory nations to stabilizing greenhouse gas concentrations in the atmosphere at a level that "would prevent dangerous anthropo-genic interference (DAI) with the climate system." In an effort to provide some insight into impacts of climate change that might be considered DAI, authors of the Third Assessment Report (TAR) ofthe Intergovernmental Panel on Climate Change (IPCC) identified 5"reasons for concern" (RFCs). Relationships between various impacts reflected in each RFC and increases in global mean temperature (GMT) were portrayed in what has come to be called the "burning embers". In presenting the "embers" in the TAR, IPCC authors did not assess whether any single RFC was more important than any other; nor did they conclude what level of impacts or what atmospheric concentrations of greenhouse gases would constitute DAI, a value judgment that would be policy prescriptive. Here, we describe revisions of the sensitivities of the RFCs to increases in GMT and a more thorough understanding of the concept of vulnerability that has evolved over the past 8 years. This is based on our expert judgment about new findings in the growing literature since the publication of the TAR in 2001, including literature that was assessed in the IPCC Fourth Assessment Report (AR4), as well as additional research published since AR4. Compared with results reported in the TAR, smaller increases in GMT are now estimated to lead to significant or substantial consequences in the framework of the 5 "reasons for concern."

**#126** Yohe, G., "Reasons for Concern" (about Climate Change) in the United States", *Climatic Change* 99: 295-302, 2010.
Article 2 of the United Nations Framework Convention on Climate Change commits its parties to stabilizing greenhouse gas concentrations in the atmosphere at a level that "would prevent dangerous anthropogenic interference with the climate system." Authors of the Third Assessment Report of the Intergovernmental Panel on Climate Change (IPCC2001a,b) offered some insight into what negotiators might consider dangerous by highlighting five "reasons for concern" (RFC's) and tracking concern against changes in global mean temperature; they illustrated their assessments in the now iconic "burning embers" diagram. The Fourth Assessment Report reaffirmed the value of plotting RFC's against temperature change (IPCC2007a,b), and Smith et al. (2009) produced an unpated embers visualization for the globe. This paper applies the same assessment and communication strategies to calibrate the comparable RFC's for the United States. It adds "National Security Concern" as a sixth RFC because many now see changes in the intensity and/or frequency of extreme events around the world as "risk enhancers" that deserve attention at the highest levels of the US policy and research communities. The US embers portrayed here suggest that: (1) US policy-makers will not discover anything really "dangerous" over the near to medium term if they consider only economic impacts that are aggregated across the entire country but that (2) they could easily uncover

"dangerous anthropogenic interference with the climate system" by focusing their attention on changes in the intensities, frequencies, and regional distributions of extreme weather events driven by climate change.

**#171** O'Neill, B.C., Oppenheimer, M., Warren, R., Hallegatte, S., Kopp, R., Portner, H., Scholes, R., Birkmann, J., Foden, W., Licher, R., Mach, K., Marbaiz, P., Mastrandrea, M., Price, J., Takahashi, K., van Ypersele, J-P., and Yohe, G., "IPCC Reasons for Concern regarding climate change risks", *Nature Climate Change* 7: 38-37, 2017.

The reasons for concern framework communicates scientific understanding about risks in relation to varying levels of climate change. The framework, now a cornerstone of the IPCC assessments, aggregates global risks into five categories as a function of global mean temperature change. We review the framework's conceptual basis and the risk judgments made in the most recent IPCC report, confirming those judgments in most cases in the light of more recent literature and identifying their limitations. We point to extensions of the framework that offer complementary climate change metrics to global mean temperature change and better account for possible changes in social and ecological system vulnerability. Further research should systematically evaluate risks under alternative scenarios of future climatic and societal conditions.

**#175** Yohe, G., "Characterizing Transient Temperature Trajectories for Assessing the Value of Achieving Alternative Temperature Targets", *Climatic Change*, 145: 469-479, 2017.

Trajectories of policy-driven transient temperatures are reported here for four different maximum temperature targets through 2100 and a no-policy baseline because it is they, and their associated manifestations in other impact and risk dimensions, that natural and human and natural systems see in real time as their common future unfolds. It follows that it is they that inform both the reactive and (for human systems) anticipatory responses that embedded decision-makers would contemplate in the future. Median pathways as well as 5thand 95th percentile alternatives for each set of scenarios are reported in decadal increments from 2010 through 2100. Two illustrations (agricultural yields and Intergovernmental Panel on Climate Change "Reasons for concern) are presented to provide provocative context within which to begin to see their potential value across a wide range of applications.

# CHAPTER 7

# DETERMINANTS OF ADAPTIVE AND MITIGATIVE CAPACITIES

*The lede – a list of underlying determinants of adaptive and mitigative capacities has been used by researchers around the world to organize their thoughts. Identifying determinants helped (#54); confirming that the "weakest link" hypothesis (#82) provided focus (with Richard Tol) is one of my most cited papers. Unintended consequences are real. It turns out that confirmed empirical work (#57) following up on insights borne of the discussions that produced #54 hampered nations' living up to their (incremental) national commitments under the United Nations Framework Convention on Climate Change (UNFCCC) toward both mitigation (#50) and adaptation (#54). Why? Because they double-counted general aid expenditures as climate contributions, and because aid contributions could improve weak determinants of capacity to respond to climate change, and so should count in the aid accounting practices. Responding to the unintended consequences of our work, this insight subsequently framed more honest international policy deliberations from Copenhagen and beyond; double counting was no longer allowed.*

This entire topic could sound like it is in the "weeds", but it was just as surely an unintended consequence that effected negotiations under the United Nations Framework Convention on Climate Change (UNFCCC) outside of the weeds. The determinants of adaptive capacity that were cited in the negotiations were derived from (#54); they included:

1. access to financial resources,
2. availability of response options,
3-5. strong human (3), social (4), and political (5) capitals,
6. a decision-making structure taking responsibility,
7. decision-makers' abilities to separate signal from noise, and
8. a population that supported all of the above.

It turns out that the determinants of mitigative capacity are essentially the same as the determinants of adaptive capacity (#50). Moreover, the parallel sets match the long-known precursors of successful public health institutions (#64 and #81); if only we and they had done our homework; see box 7-1.

On the basis of this coincidence, I wrote that support for improving human, social, and political capital and statistical training would be good climate policy. It was then, and it is still. I never thought, however, that this conclusion would be used by countries like the United States in UNFCCC negotiations to support government claims that "we are already supporting climate policy" because we provide "this (or that) in aid for improved government functions or educating children" and we have been doing so for many years. The result was nations providing nothing new of in overall international aid for many years.

I worked with Richard Tol to publish a paper on a "weakest link" hypothesis for adaptive capacity (#82) – the idea was that the adaptive capacity (and/or the mitigative capacity by analogy) of a country or community is fundamentally determined by the weakest of the underlying determinants. We suggested a way to implement this is to organize humanity's thinking about how to frame policy for decision-makers who always faced a scarcity of resources. The hypothesis has been confirmed widely. It is still being widely cited and more honestly applied (more than 1500 citations and counting).

Looking at the underlying determinants, I wrote that the US was strong in adaptive capacity and weak in mitigative capacity because the costs and benefits of the two responses were differently distributed (#50). In the US, public investment in adaptation spread the cost widely for the benefit of a few well-connected victims of climate change – high adaptive capacity. Public mitigation, by contrast, would spread the cost of emissions reductions on a few well-connected energy companies while the victims who would benefit were distributed widely across the globe and well into the future – low mitigative capacity.

Richard Schmalensee had already taught me that political economy was critical. Calculating quantitatively or surmising qualitatively played well into explaining the political economy – the US sometimes adapts well, but it seems always reluctant to mitigate.

Happily, the evidence for low mitigative capacity weakened after the 2020 election. Corporations across the country had already been committing to accomplishing their shares of the Paris Agreement mitigation targets even without leadership from D.C. Why? Not because of concern about the climate, but because doing so would be good for the bottom line. They understood that the world was moving to price carbon (or at least place a shadow price on carbon in a permit market) at a rate that would grow over time, even without the United States' participation in the Paris Accord. The US's returning to the Accord in 2021 made the signal stronger, but international negotiators were not convinced that the US would remain committed.

## References, abstracts and links connected to my January 2023 CV

**#50** Yohe, G., "Mitigative Capacity – The Mirror Image of Adaptive Capacity on the Emissions Side", *Climatic Change* 49: 247-262, 2001.
Two principles upon which productive exploration of this diversity might be organized have begun to emerge in the collective psyche of analysts across the globe. One, adaptive capacity, emerged over the past two years on the impacts side of the climate equation; and it has already been used to provide general insight from specific examples, to highlight opportunities for diminishing climate related damage, and to organize research activity. The second, mitigative capacity, is the mirror image of adaptive capacity on the emissions side of the equation. Mitigative capacity is a newer concept; but it is the intent of this editorial to argue that it, too, holds the promise of offering instructive lessons and focused hypotheses.

**#54** Section 18.5 in Smit, B., Pilifosova, O., Burton, I., Challenger, B., Huq, S., Klein, R.J.T., and Yohe, G., "Adaptation to Climate Change in the Context of Sustainable Development and Equity", in *Climate Change 2001: Impacts, Adaptation and Vulnerability,* Cambridge: Cambridge University Press, 2001.
Adaptation to climate change and risks takes place in dynamic social, economic, technological, biophysical, and political contexts that vary over time, location, and sector. This complex mix of conditions determines the capacity of systems to adapt. Although scholarship on adaptive capacity is extremely limited in the climate change field, there is considerable understanding of the conditions that influence the adaptability of societies to climate stimuli in the fields of hazards, resource management, and sustainable development. From this literature, it is possible to identify the main features of communities or regions that seem to determine their adaptive capacity: economic wealth, technology, information and skills, infrastructure, institutions, and equity.

**#57** Yohe, G. and Tol, R., "Indicators for Social and Economic Coping Capacity – Moving Toward a Working Definition of Adaptive Capacity", *Global Environmental Change* 12: 25-40, 2002.

This paper offers a practically motivated method for evaluating systems' abilities to handle external stress. The method is designed to assess the potential contributions of various adaptation options to improving systems' coping capacities by focusing attention directly on the underlying determinants of adaptive capacity. The method should be sufficiently flexible to accommodate diverse applications whose contexts are location specific and path dependent without imposing the straightjacket constraints of a "one size fits all" cookbook approach. Nonetheless, the method should produce unitless indicators that can be employed to judge the relative vulnerabilities of diverse systems to multiple stresses and to their potential interactions. An artificial application is employed to describe the development of the method and to illustrate how it might be applied. Some empirical evidence is offered to underscore the significance of the determinants of adaptive capacity in determining vulnerability; these are the determinants upon which the method is constructed. The method is, finally, applied directly to expert judgments of six different adaptations that could reduce vulnerability in the Netherlands to increased flooding along the Rhine River.

**#64** Yohe, G. and Ebi, K. "Approaching Adaptation: Parallels and Contrasts between the Climate and Health Communities" in *Integration of Public Health with Adaptation to Climate Change: Lessons Learned and New Directions* (Ebi, K., Smith, J. and Burton, I., eds.), Taylor and Francis, The Netherlands, 2005. The content is obvious from the title/

**#81** Tol, R., Ebi, K., and Yohe, G., "Infectious Disease, Development and Climate Change: A Scenario Analysis", *Environment and Development Economics* 12: 687-706, 2007.

We study the effects of development and climate change on infectious diseases in Sub-Saharan Africa. Infant mortality and infectious disease are closely related, but there are better data for the former. In an international cross-section, per capita income, literacy, and absolute poverty significantly affect infant mortality. We use scenarios of these three determinants and of climate change to project the future incidence of malaria, assuming it to change proportionally to infant mortality. Malaria deaths will first increase, because of population growth and climate change, but then fall, because of development. This pattern is robust to the choice of scenario, parameters, and starting conditions; and it holds for diarrhoea, schistosomiasis, and dengue fever as well. However, the timing and level of the mortality peak is very sensitive to assumptions. Climate change is important in the medium term, but dominated in the long term by development. As climate can only be changed with a substantial delay, development is the preferred strategy to reduce infectious diseases even if they are exacerbated by climate change. Development can, in particular, support the needed strengthening of disease control programs in the short run and thereby increase the capacity to cope with projected increases in infectious diseases over the medium to long term. This conclusion must, however, be viewed with caution.

**#82** Tol, R. and Yohe, G., "The Weakest Link Hypothesis for Adaptive Capacity: An Empirical Test", *Global Environmental Change* 17: 218-227, 2007.

Yohe and Tol (2002. Global Environmental Change 12, 25–40) built an indexing method for vulnerability based on the hypothesis that the adaptive capacity for any system facing a vector of external stresses could be explained by the weakest of its underlying determinants—the so-called "weakest link" hypothesis. Their structure noted eight determinants, but the approach could handle any number. They quoted analogies in support of the hypothesis, but loose inference is hardly sufficient to confirm such a claim. We respond to this omission by offering an empirical investigation of its validity. We estimate a structural form designed to accommodate the full range of possible interactions across sets of underlying determinants. The perfect complement case of the pure "weakest-link" formulation lies on one extreme, and the perfect substitute case where each determinant can compensate for all others at constant rates is the other limiting case. For vulnerability to natural disasters, infant mortality and drinking water treatment, we find qualified support for a modified weakest link hypothesis: the weakest indicator plays an important role because other factors can compensate (with increasing difficulty). For life expectancy, sanitation and nutrition, we find a relationship that is close to linear— the perfect substitute case where the various determinants of adaptive capacity can compensate for each other with relative and persistent ease. Moreover, since the factors from which systems derive their adaptive capacities are different for different risks, we have identified another source of diversity in the assessment of vulnerability.

## Box 7-1 A second exploration of the parallels between analyses of COVID-19 and climate change

On February 1$^{st}$ of 2022, *Lancet* published a paper from more than 100 authors of the COVID-19 National Preparedness Collaborators (NPC). They reconfirmed that analysis of COVID shines a light onto the character of societies around the world as they rise to meet the challenges of a significant external force. Moreover, their work contributes to another literature – the ensemble of parallel analyses of how humanity might respond to the existential risks born of climate change, and that is the point of this correspondence.

It had become noticeably evident over the past several years that time dimensions of drivers and impacts were the only fundamental differences between analyses of a global pandemic like COVID-19 and the global threat of climate change. For present purposes, call that statement *Hypothesis #1*. It is anchored on two parallel practical facts: on the one hand, the mortality and morbidity risks of a pandemic put humanity in harm's way along time scales that were measured in days (e.g., 7-day intervals), months, and years; but human responses were designed to reduce those risks by either lowering the likelihood of an individuals' being infected or the lessening the consequences of such an infection. In contrast, the human and economic risks of climate change put humanity and its posterity at risk along scales that were measured in seasons, 5- or 10-year trends, decades, and centuries; and current responses to a very long-term problem were designed to reduce either the likelihood of specified levels of warming or the consequences of coping with the resulting impacts. To summarize, the choices in either context are three: abate (mitigate in the climate jargon), adapt, or suffer.

Thinking about the adaptation choice in the climate change arena has been formalized by expressing vulnerability to climate risks (V) as a potentially complex and site-specific function of exposure (E) and sensitivity (S). Both of these vectors are themselves seen to be functions of adaptive capacity (*AC*) – an invention of the authors of chapter 18 ("Adaptation") in the contribution of Working Group II to the Third Assessment Report of the Intergovernmental Panel on Climate Change.[1] It was appropriate to assume that V would increase with an increase in any component of the vector E, while it would decline with at an increasing rate with the increase of any component of vector S. These can be *Hypotheses #2 and #3* for present purposes.

In its formal structure, AC was taken to be a secondary function of N site-specific and not-necessarily-independent multivariate "determinants" ($D_1$, ...., $D_N$):

$$V = f\{E(AC); S(AC)\} \text{ where } AC = g\{D_1; ....; D_N\}.$$

An up-to-date version of the established determinants of the adaptive capacity (and the analogous mitigative capacity) of societies to respond to an external stress can include:

$D_1$ – the availability to response options available to society, including risk spreading mechanisms,

$D_2$ – the availability to resources and the character of their distributions across the relevant population,

$D_3$ – the strength and credibility of society's critical decision- and opinion-making institutions and their decision criteria,

$D_4$ – the stock of human capital across the population, including educational achievement and personal security,

$D_5$ – the stock of social, political-economic and legal capital across the society,

$D_6$ – the ability of society's decision- and opinion-makers to comprehend, manage, and communicate dynamic sources of evolving information and maintain their credibility across the population, and

$D_7$ – the public's perception of the sources of the external stressors and the significance of that stress in determining exposure and sensitivity to their manifestations.

This thread in the literature quickly expanded in its real-world applicability by hypothesizing that the overall capacity to adapt to (or mitigate against) the manifestations of an external stress would depend most significantly on the weakest of the underlying determinants. Let this be *Hypothesis #4*. Many have nonetheless argued that this formalization implies that policies and actions to strengthen any determinant $D_i$ can be viewed as potentially effective climate or health policy.

Turning now to the value of NPC, it is productive to apply the formalization of adaptive capacity with respect to climate risks across the interface. Quite simply, "infections per capita" can be thought of as exposure E, while the "infection fatality ratio" (IFR) is a reflection of sensitivity S.

To see this point, notice that the explanatory variables for variation in E and S across the cross-sectional data that has been analyzed by the NPC can be easily placed within the boundaries of one or more of the determinants listed in the climate context:

- From their Table 2, for example, "GDP per capita" fits inside $D_2$, "trust in government" touches base with $D_7$, while "interpersonal trust" reflects $D_4$.
- In their Table 3, "pandemic preparedness" indicators belong to $D_1$ and $D_3$, "health care capacity" indicators populate $D_1$ and $D_2$ , "governance indicators" link to $D_3$, $D_5$, and $D_7$, while "social indicators" fit into $D_2$, $D_4$, $D_6$, and $D_7$.
- Their Figure 4 correlates elements of the $D_1$ vector (where response options include mobility and vaccination) with the three significant indicators of $D_7$ ("trust in government"), $D_4$ ("interpersonal trust"), and $D_5$ ("government corruption").

It would appear that the authors of the NPC could have organized their analysis and their communication of results in ways that are entirely consistent with *Hypotheses #1.*

Consistent with the implications of *Hypotheses #2 and #3*, their Figure 2 clearly shows a strong negative correlation between E and S, especially when the months between October 15, 2020 and September 30, 2021 are included.

Finally, the major conclusions reported by the NPC authors bear witness to the applicability of the "weakest link" interpretation of Hypothesis #4:

1. Policy-makers (and opinion-makers, one would expect) cannot influence many of contextual factors that explain variation in both E and S.
2. Important indicators used to construct healthcare capacity and preparedness indices are not correlated in cross-sectional variation in either E and S.
3. Trust in government and other people are significant in explaining variation in E but not S.
4. Vaccinations may be the mechanism behind the significance of the trust conclusions.

*Trust is the weakest link in the pandemic context* – that is, the NPC has shown that lack of trust is the primary obstacle to reducing health risk from COVID-19 (by lowering likelihood of exposure and/or is consequences through contextual factors), and that is alarming.

Confidence in this finding should be very high because it was produced from very careful, copious, and comprehensive analysis of an existential health threat for all of humanity. By virtue of the strength of the COVID-climate analogy described here and this high confidence, the NPC's work also casts some significant shade on the hope that climate change interventions will work well without our (somehow) rebuilding of trust in our understanding of the climate-socioeconomic-political system across public and private sectors.

It follows from the analogy at the interface that methods that improve trust in either domain should benefit both, because they would help strengthen the weakest link in both response domains. Two corollaries also emerge.

First, *hypothesis #1* is another reason why improved collaboration across the health-climate interface would be valuable in developing and applying modeling and measurement techniques in support of ensemble projections and counterfactual exercises. Secondly, the strengths of *hypotheses #2 to #4* mean that increased collaboration across climate and epidemiologic boundaries on ways transparently to preserve and communicate the credibility of the science will do the same in support of productive public discourse.

# CHAPTER 8

# THE IPCC AND THE MEANING OF CONSENSUS

*The lede: before going back into the weeds about iterative risk management in chapter 9, it is important to cover and hopefully correct one large, publicly held misconception. It turns out that the applicable concept of consensus is completely misunderstood across the collection of casual observers and denying critics of the climate problem. Consensus does not mean that everyone in the room votes "yes" – word by word and number by number. Instead, it means that nobody votes "no" about anything.*

The rules for IPCC plenary meetings, where reports are approved, and the annual Conferences of the Parties (COPs) of the United Nations Convention on Climate Change (UNFCCC) abide by the definition of consensus that is employed in most if not all international negotiations. It applies to the Summaries for Policymakers at IPCC plenaries, and it applies to any proposed finding, statement, or action item emerging from a COP.

The process is this, using the IPCC plenaries as the example: each sentence of the Summary for Policymakers is put before the assembly of representatives from the now 197 signatory nations who have ratified the UNFCCC. That is their ticket to gain access to the meetings. Every paragraph is taken one at a time, as is each sentence. Each graph or table is also taken one at a time. Anyone in the room from any country can object to any word in the sentence before them. Anyone can object to any number on any table or any line on any graph. If that happens, the entire room works on that word or number or line until nobody in the room objects to whatever revision has emerged from the process. Then, the room can move on to considering the next word or line or number. Alternatively, the plenary can {bracket} a contentious item for later; but it will always come back around for consensus after rounds of formal and informal sidebar discussions and group meetings to find common ground.

That is why it takes so long to approve an IPCC report. That is also why every nation buys into the product for subsequent negotiations about what to do. They know that, under the rules of the UNFCCC, whatever language emerges from IPCC plenaries will be the basis of negotiations for up to six years.

IPCC is bound to this process by its charter. It is a challenge, but it works. That is why the Third National Climate Assessment (NCA3) in the United States (and all subsequent NCAs) accepted the same approach to achieving consensus. The NCA3 version required that any author in the room who objected to any word (or line on a graph or entry in a table, etc...) had to begin the negotiating process by suggesting and defending an alternative. When that happened, their suggestion would immediately become the topic on the table for discussion. Iteration from word to word could take hours, and many times it would end back where it started (i.e., the original language). Still, the end result was consensus on the content of any particular conclusion and evidence that all voices had been heard. To invent another deliberate double negative: consensus means that "nobody disagreed".

To be clear, and this is important, consensus does not mean that nobody in the room disagreed with a conclusion that "climate change would do this, or that". The language to which someone had objected would generally have included a confidence statement. This is where much of the negotiation would focus. So, the plenary decisions in these setting are asserting that nobody disagreed with a statement like "with x degree of confidence, this or that conclusion that something would be a manifestation of climate change that had been detected and its connection to climate change had been attributed." The confidence language as described in chapter 5 helps a lot.

Assessments' reporting of high risk from any calculation now simply had to display significant value added in terms of likelihood and consequence and provide credibility for decision-makers who understood risk even if it could not be quantified (see #s 69, 70, 84, 85, 86, 111, 113, 114, 115 and 117). "Reasons for Concern" (see chapter 6) are a perfect example; they were invented in the Third Assessment Report and the judgements are subjective. It is essential to understand that, from the TAR through the AR6, their content achieved consensus status through both the language of the findings and their portrayal in the "burning embers" (except in the AR4).

# References, abstracts and links connected to my January 2023 CV

**#71** Tol, R. and Yohe, G., "On Dangerous Climate Change and Dangerous Emission Reduction", in *Avoiding Dangerous Climate Change* (Schellnhuber, H.J., Cramer, W. Nakicenovic, N. Wigley, T., and Yohe, G. eds.), Cambridge: Cambridge University Press, 2006.

The International Symposium on Stabilization of Greenhouse Gas Concentrations, Avoiding Dangerous Climate Change, (ADCC) took place, at the invitation of the British Prime Minister Tony Blair and under the sponsorship of the UK Department for Environment, Food and Rural Affairs (Defra), at the Met Office, Exeter, United Kingdom, on 1–3 February 2005.

The conference attracted over 200 participants from some 30 countries. These were mainly scientists, and representatives from international organisations and national governments. The conference offered a unique opportunity for the scientists to exchange and debate their views on the consequences and risks presented to the natural and human systems as a result of changes in the world's climate, and on the pathways and technologies to limit GHG emissions and atmospheric concentrations. The conference took as read the conclusions of the IPCC Third Assessment Report (TAR) that climate change due to human actions is already happening, and that without actions designed to reduce emissions climate will continue to change with increasingly adverse effects on the environment and human society.

**#72** Yohe, G., Adger, N., Dowlatabadi, H., Ebi, K., Huq, S., Moran, D., Rothman, D., Strzepek, K., and Ziervogel, G., "Recognizing Uncertainty in Evaluating Responses", in *Ecosystems and Human Wellbeing: Volume 3 – Policy Responses,* New York: Island Press, 2006.

Decision-makers face pervasive uncertainty in implementing response strategies as they try to manage ecosystem services. Uncertainty clouds their understanding of everything from how their response options might actually work to the methods that they use to assess their relative efficacy. This chapter takes this simple observation as a point of departure and tries to provide an insight into how valuation and decision-analytic frameworks can accommodate uncertainty. It also offers some guidance to those who want to assess how uncertainty combines with issues of political feasibility and governance (discussed in Chapter 3) to affect the confidence with which they can trust their conclusions about how best to respond. Both objectives recognize the fundamental truth that decision-makers have to make decisions even when uncertainty is extremely large; and both recognize that maintaining the status quo (that is, enacting no new response to one or more new sources of stress) is as much of a decision as moving robustly in many directions at the same time.

**#88** Janetos, A., Balstad, R., Apt, J., Ardanuy, P, Friedl, R., Goodchild, M., Macauley, M., McBean, G., Skole, D., Welling, L., Wilbanks, T., and Yohe, G., "Earth Science Applications to Societal Benefits", in *Earth Science and Applications from Space: National Imperatives for the Next Decade and Beyond*, Washington, D.C.: National Research Council, The National Academies Press, 2007.

Increasing the societal benefits of Earth science research is high on the priority list of federal science agencies and policy makers, who have long believed that the role of scientific research is not only to expand knowledge but also to improve people's lives. Although promoting societal benefits and applications from basic research has been emphasized in national science policy discussions for decades, policy and decision makers at federal, state, and local levels also increasingly recognize the value of evidence-based policy making, which draws on scientific findings and understandings. The theme of this chapter is the urgency of developing useful applications and enhancing benefits to society from the nation's investment in Earth science research.

Accomplishing this objective requires an understanding of the entire research-to-applications chain, which includes generating scientific observations, conducting research, transforming the results into useful information, and distributing the information in a form that meets the requirements of both public and private sector managers, decision makers, policy makers, and the public at large.

There are a number of remarkable successes in reaping the benefits of Earth science research. For example, this study documents that many nations of the developed world have created sophisticated flood forecast systems that use precipitation gauges and radars, river stage monitoring, and weather prediction models to create warnings of floods from hours to several days in advance. However, there is no global capacity to do this, and developing nations are largely without this capability. As a response to this concern, the Panel on Earth Science Applications and Societal Benefits offers observations on how to move from discovery to design, balance mission portfolios to benefit both research and applications, and establish mechanisms for including the priorities of the applications community in space-based measurements.

**#102** Yohe, G., "Inside the Climate Change Panel: Babel of Voices, a Single Conviction", *The InterDependent* 5: 14, Winter 2007/2008, cited in https://www.ipcc.ch/site/assets/uploads/2018/03/inf2-7.pdf

**#117** Smith, J.B., Schneider, S. H., Oppenheimer, M., Yohe, G., Hare, W., Mastrandrea, M.D., Patwardhan, A., Burton, I., Corfee-Morlot, J., Magadza, C.H.D., Füssel, H-M, Pittock, A.B., Rahman, A., Suarez, A., and van Ypersele, J-P, "Dangerous Climate Change: An Update of the IPCC Reasons for Concern", *Proceedings of the National Academy of Science* 106: 4133-4137, March 17, 2009.

The UNFCCC (http://unfccc.int/resource/docs/convkp/conveng.pdf) commits signatory nations to stabilizing greenhouse gas concentrations in the atmosphere at a level that "would prevent dangerous anthropo-genic interference (DAI) with the climate system." In an effort to provide some insight into impacts of climate change that might be considered DAI, authors of the Third Assessment Report (TAR) of the Intergovernmental Panel on Climate Change (IPCC) identified 5"reasons for concern" (RFCs). Relationships between various impacts reflected in each RFC and increases in global mean temperature (GMT) were portrayed in what has come to be called the "burning embers diagram." In presenting the "embers" in the TAR,IPCC authors did not assess whether any single RFC was more important than any other; nor did they identify what level of impacts or what concentrations of greenhouse gases would constitute DAI. A value judgment like that would be policy prescriptive. Here, we describe revisions of the sensitivities of the RFCs to increases in GMT and a more thorough understanding of the concept of vulnerability that has evolved over the past 8 years. This is based on our expert judgment about new findings in the growing literature since the publication of the TAR in 2001, including literature that was assessed in the IPCC Fourth Assessment Report (AR4), as well as additional research published sinceAR4. Compared with results reported in the TAR, smaller increases in GMT are now estimated to lead to significant or substantial consequences in the framework of the 5 RFCs.

**#119** Yohe, G. "Addressing Climate Change through a Risk Management Lens", in *Assessing the Benefits of Avoided Climate Change: Cost Benefit Analysis and Beyond*, (Gulledge, J., Richardson, L., Adkins, L., and Seidel, S., eds.), *Proceedings of Workshop on Assessing Benefits of Avoided Climate Change, March 16–17, 2009*, Arlington, VA.: Pew Center on Global Climate Change, 2009, p. 201–231.

In the Summary for Policymakers of the Synthesis Report for its Fourth Assessment, the Intergovernmental Panel on Climate Change (IPCC) achieved unanimous agreement from signatory countries of the Framework Convention that, "Responding to climate change involves an iterative risk management process that includes both adaptation and mitigation, and takes into account climate change damages, cobenefits, sustainability, equity and attitudes to risk". By accepting this key sentence, governments recognized for the first time, that their negotiations and associated policy deliberations must, individually and collectively, be informed by views of the climate problem drawn through the lens of reducing risk.

As new as this perspective might be for climate policy negotiators, risk-management is already widely used by policymakers in other decision making processes, such as designing social safety net programs, monetary policy, and foreign policy. Even though governments and some segments of the policy community are comfortable with the risk management paradigm, however, the climate change research and assessment community had heretofore been slow to catch on.

This paper presents a first attempt to deconstruct the application of a risk-based paradigm to climate change by considering the critical phrases that are highlighted above, offering insights into what we do and do not know in each case.

Perhaps most importantly, the typical cost-benefit analysis used to make decisions in establishing regulations may not be fully appropriate for the climate problem because, to a large degree, many damages cannot be expressed monetarily and because uncertainty is so pervasive. To avoid being hamstrung by these fundamental complications, traditional policy analyses need to be supplemented by risk-based explorations that can more appropriately handle low-probability events and more easily handle large consequences calibrated in non-monetary metrics. In short, adopting a risk-based perspective will bring new clarity to our understanding of the diversity and complexity of the climate problem.

# CHAPTER 9

# ITERATIVE RISK MANAGEMENT

*The lede – this also sounds like it is in the weeds, but it is NOT. These three words are my most important contribution to the planet. In 2007, they had concluded that our detection of significant warming was unequivocal. Authors also had very high confidence that human emissions of greenhouse gases were the primary cause. Deniers had moved from "the world is not warming" and "human activity has not played a role" to asking with straight faces "So what? impacts are negligible".*

*Mentioning extremes in the distribution of any of the Reasons for Concern was, to their minds and their polemics, divisive at best, and shameless "fear-mongering" at worst. For sure, they would argue, it was partisan, because only Democrats could ignore the catastrophic and obvious conspiracy among climate change researchers to pad their pockets by preserving their funding. The idea was that we all talk among ourselves and agree on a few false talking points about how the world operates so that we can protect our way of life. They told the world that scientists travel to IPCC meetings on private jets. They stayed in the best hotels at the taxpayers' expense; and they otherwise distracted humanity from the truth.*

*"Look", they would say, "carbon dioxide is as good for the trees and the bees as it is for the crops that feed us. What could possibly be the harm of having a little bit more? All that would do produce more food. And the science is a scam. How can a miniscule increase in concentration of something as positive as carbon dioxide possibly cause significant damage? Or any damage at all, for that matter? All this stuff has been happening for centuries, but NOW it is a problem? You scientists don't possibly have enough information to weigh the damages you invent from God's gift against the costs that society will most certainly endure if you have your way."*

As noted above, and below, the IPCC AR4 Summary for Policymakers of the Synthesis Report (#85) included, on page 22, the following words: "Responding to climate change involves an ***iterative risk management*** process that includes both *adaptation* and *mitigation* and takes into account climate change *damages, co-benefits, sustainability, equity, and attitudes to risk*" (bold italics are my emphasis). From that point on, after it was approved by all signatory countries of the United Nations Framework Convention on Climate Change (UNFCCC), scientists' reporting about extremes was responding to the needs of our clients – the nations of the world.

These 30 words were crafted during the Synthesis Report authors' meeting in Estes Park, Colorado - early in the mornings, after many long nights by three of us: Stephen Schneider, William Hare, and myself, with the strong support from IPCC Chair Rajendra Pachauri. We had an idea that those words could certainly change the way that decision-makers and opinion-makers across the planet looked at climate change as a policy issue. Cost-benefit analysis would no longer be the only tool policy design tool. Risk management was the way forward.

To be clear, after the "iterative risk management" language had been approved by consensus in 2007, IPCC plenary meetings would allow "x degree of confidence" to be very small if the consequences of the manifestations could be very large –that is what the client nations wanted to know, because risk is likelihood multiplied by consequence.

Understanding that IPCC and NCA conclusions had achieved this level of consensus was essential. It gave extra credibility to results from other assessments, like America's Climate Choices, Risky Business, and the New York Panel on Climate Change, that used the same process. IPCC and the NCA (National Climate Assessment for the United States - #3 in 2014 and #4 in 2018) are not policy-prescriptive. America's Climate Choices (ACC 2010 from the National Academies of Science) and NPCC (New York Panel on Climate Change - #1, #2, and #3 in 2008, 2012 and 2018) were created, though, by request of their sponsors (the sponsor of ACC was the United States Congress).

To emphasize my point, accepting an "iterative risk management" approach meant that the clients of assessments from whatever source wanted authors to report low confidence possibilities if they could describe potentially high consequences, because low likelihood times high consequence could mean high risk. This led to reporting what could happen in the dark tails of climate futures without embarrassment and without vulnerability to claims of fear-mongering.

It took years for some to get the message. It took others like Michael Bloomberg fewer than 60 seconds. When the word was passed to the Mayor on the elevator trip from the basement to the fifth floor in response to an intervention from the New York (City) Panel on Climate Change by his chief of staff, Adam Freed, the response was something like "this is a 'no brainer'" from a man who had made many fortunes managing risk. "Make it happen" was his order.

Of course, he is not the only one who knows how to manage risk. Most of us buy home and car insurance even if it is not required. We invite dogs into our homes for protection even if all they will do is lick the face of an intruder. We install alarm systems in case that our small phu-phu dogs will just love up to an intruder. We build dikes even though we know that they will not necessarily eliminate the chance of harm in the extremes. We all live with risk, and we all know that we cannot eliminate it – yet we all know that it is worthwhile to try to achieve something "tolerable"..

I personally thought at the time (in 2007), and I still think in 2023 (and I am not alone), that these words are perhaps among the most significant conclusion of the six IPCC assessments that have been produced so far. Steve Schneider agreed, and sacrificed his health trying to sell the message until his passing on an airplane flying from one talk in Scandinavia to another in the UK.

I received word simultaneously with colleagues having breakfast at the River Inn preparing for another Academy meeting.

Steve, Bill, and I worked to frame this language in Estes Park. We also worked, with Pachauri, in Valencia to defend it. Pachauri brilliantly put those words on the calendar for the morning of day one of the IPCC plenary in front of the then-169 countries who had ratified the UNFCCC by that time. The idea was that the world would have as much time as possible to meet with admirers and detractors of each and every word to hammer out some satisfactory language – word for word, by consensus.

The United States pushed back, but their delegation was met in an offline discussion in a room populated by Pachauri (the host), Schneider, Hare, and myself. The United States delegation was expecting to see Steve in the room, but not the rest of us... "We did not know that more than one world respected author would be here" (Ko Barrett said those words). It was a short meeting; the US agreed.

It turned out, after multiple iterations, that all thirty of the original words (from Estes Park) ultimately achieved consensus approval on the last afternoon of the plenary – five days of negotiation brought us back to where we started. Go figure!

Please understand that, moving forward, these words were the motivation of the Paris Agreement in December of 2015, and they depend, neither in the US nor across the globe, on leadership from Washington.

Please also understand that these words do not expel cost-benefit approaches to adaptation (#65 discusses how decisions can be C-B in the short-run if they are designed to be reactive depending upon detection in the long-run).

In the long-run, though, attribution to human activity comes into play, so that only wide ranges of projections are possible. In this decision environment, risk management is the correct lens. It has become the standard across the country and around the world.

So as not to leave any confusion, here are a few additional basic insights from the myriad of papers noted above and from the language that was accepted:

1. We cannot write policy for 100 years, so we have to iterate; but that is not new news to corporations and communities who can respond in the medium term based on risk-based mid-course corrections.
2. Risk is likelihood times consequence – risk matrices allow for qualitative or quantitative evaluation; and "tolerable risk" is a working example of an appropriate version of the precautionary principle.
3. The final message: do not plan for the worst, but do consider something in the upper extremes.

Numbers 42, 100, 105, 118, 120, and 152 provide some insight into just how hard of a climb that can be.

# References, abstracts and links connected to my January 2023 CV

**#42** Yohe, G., "The Tolerable Window Approach - Lessons and Limitations", *Climatic Change* 41: 283-295, 1999.
The Tolerable Windows Approach (TWA) described in Petschel-Held et al. (1999) is one representative of a new approach to analyzing global change mitigation policy. The TWA attempts to define the boundaries of tolerable change that might serve as guardrails against catastrophic impacts. It then tries to work 'backwards' to see how we might constrain the emission of greenhouse gases to guarantee that those boundaries are never crossed and the associated guardrails are never tested. As such, the TWA joins at least three other analytical approaches that have been exercised over the past decade or so to examine mitigation policy. Each approach has its own strengths and its own weaknesses; but the strengths of some have tended to complement the weaknesses of the others so that their combined contribution to our understanding of global change has been greater than the sum of their individual contributions. Armed with insight into the value of multiple approaches, both the research community and its constituent collection of policy-types have welcomed TWA into the analytical fold.

**#43** Yohe, G. and Dowlatabadi, H., "Risk and Uncertainties, Analysis and Evaluation: Lessons for Adaptation and Integration", *Mitigation and Adaptation Strategies for Global Change* 4: 319-329, 1999.
This paper draws ten lessons from analyses of adaptation to climate change under conditions of risk and uncertainty: (1) Socio-economic systems will likely respond most to extreme realizations of climate change. (2) Systems have been responding to variations in climate for centuries. (3) Future change will affect future citizens and their institutions. (4) Human systems can be the sources of surprise. (5) Perceptions of risk depend upon welfare valuations that depend upon expectations. (6) Adaptive decisions will be made in response to climate change and climate change policy. (7) Analysis of adaptive decisions should recognize the second-best context of those decisions. (8) Climate change offers opportunity as well as risk. (9) All plausible futures should be explored. (10) Multiple methodological approaches should be accommodated. These lessons support two pieces of advice for the Third Assessment Report: (1) Work toward consensus, but not at the expense of thorough examination and reporting of the "tails" of the distributions of the future. (2) Integrated assessment is only one unifying methodology; others that can better accommodate those tails should be encouraged and embraced.

**#101** Yohe, G., "A Reason for Optimism" *Tiempo Climate Newswatch*, January 2008.
In any case, governments have, in this simple but profound change in attitude, finally asked the authors of IPCC assessments to provide information about "climate risks". These are the governments that have signed on to the climate treaty, the UNFCCC. These are the governments that negotiate global climate policy. These are the governments who hold the future of the planet in their hands. These are the

governments that now understand that they have, heretofore, been asking the wrong questions.

**#61** Yohe, G., Andronova, N. and Schlesinger, M., "To Hedge or Not Against an Uncertain Climate Future", *Science*, 306: 415-417, October 15, 2004.
It has been over a decade since Nordhaus (1) published his seminal paper on mitigation policy for climate change. His question was "To slow or not to slow?"; his answer was derived from a traditional cost-benefit approach. He found that a tax levied on fossil fuel in proportion to its carbon content, which would climb over time at roughly the rate of interest, maximized global welfare. Although many more analyses of the same question have since been published, his results are still robust if one assumes a deterministic world in which decision-makers are prescient. However, no decision-maker has perfect foresight, and the uncertainty that clouds our view of the future has led some to argue that near-term mitigation of greenhouse gas emissions would be foolish. Such policy would impose immediate costs, they argue, and have un-certain long-term benefits. We take a different approach

**#63** Yohe, G., "Some Thoughts on Perspective", *Global Environmental Change* 14: 283-286, 2004.
This issue of Global Environmental Change (and the OECD Workshop from which it emerged) focuses our attention on the benefits of climate policy—specifically, the damages of climate change and climate variability that could be avoided by mitigation. Having some idea of the size of the benefits that can be attributed to mitigation would certainly seem to be a critical piece of information for decision-makers who are contemplating global responses to climate change from a cost benefit perspective. Notwithstanding the contributions of the authors whose work appears in this issue and elsewhere, however, we should be concerned that the research community has not yet advanced to the point where it can offer these decision-makers reliable estimates of global benefits that they require. Building on the insights reported by Yohe and Schlesinger (2002), this note explains why in Section 1 it is essential to recognize that these concerns do not imply that decision-makers should delay mitigation interventions until our knowledge of global benefits improves. Instead, as argued in Section 2, these concerns suggest that a different, risk-based precautionary perspective might be a more appropriate context within which to frame discussions.

**#105** Keller, K., Yohe, G., Schlesinger, M., "Managing the Risks of Climate Thresholds: Uncertainties and Information Needs", *Climatic Change* 91: 5-10, 2008.
Human activities are driving atmospheric greenhouse-gas concentrations beyond levels experienced by previous civilizations. The uncertainty surrounding our understanding of the resulting climate change poses nontrivial challenges for the design and implementation of strategies to manage the associated risks. One challenge stems from the fact that the climate system can react abruptly and with only subtle warning signs before climate thresholds have been crossed. Model predictions suggest that anthropogenic greenhouse-gas emissions increase the likelihood of crossing these thresholds. Coping with deep uncertainty in our under-

standing of the mechanisms, locations, and impacts of climate thresholds presents another challenge. Deep uncertainty presents itself when the relevant range of systems models and the associated probability density functions for their parameterizations are unknown and/or when decision-makers strongly disagree on their formulations. Further-more, the requirements for creating feasible observation and modeling systems that could deliver confident and timely prediction of impending threshold crossings are mostly un-known. These challenges put a new emphasis on the analysis, design, and implementation of Earth observation systems and strategies to manage the risks of potential climate threshold responses.

**#118** Anthoff, D., Tol, R.S.J, and Yohe, G., "Discounting for Climate Change", *Economics: The Open-Access, Open-Assessment E-Journal*, **3**: 2009-24, 2009.
It is well-known that the discount rate is crucially important for estimating the social cost of carbon, a standard indicator for the seriousness of climate change and desirable level of climate policy. The Ramsey equation for the discount rate has three components: the pure rate of time preference, a measure of relative risk aversion, and the rate of growth of per capita consumption. Much of the attention on the appropriate discount rate for long-term environmental problems has focused on the role played by the pure rate of time preference in this formulation. We show that the other two elements are numerically just as important in considerations of anthropogenic climate change. The elasticity of the marginal utility with respect to consumption is particularly important because it assumes three roles: consumption smoothing over time, risk aversion, and inequity aversion. Given the large uncertainties about climate change and widely asymmetric impacts, the assumed rates of risk and inequity aversion can be expected to play significant roles. The consumption growth rate plays multiple roles, as well. It is one of the determinants of the discount rate, and one of the drivers of emissions and hence climate change. We also find that the impacts of climate change grow slower than income, so the effective discount rate is higher than the real discount rate. Moreover, the differential growth rate between rich and poor countries determines the time evolution of the size of the equity weights. As there are a number of crucial but uncertain parameters, it is no surprise that one can obtain almost any estimate of the social cost of carbon. We even show that, for a low pure rate of time preference, the estimate of the social cost of carbon is indeed arbitrary—as one can exclude neither large positive nor large negative impacts in the very long run. However, if we probabilistically constrain the parameters to values that are implied by observed behavior, we find that the expected social cost of carbon, corrected for uncertainty and inequity, is approximate 60 US dollar per metric tonne of carbon (or roughly $17 per tonne of $CO_2$) under the assumption that catastrophic risk is zero.

**#120** Yohe, G., "Toward an Integrated Framework Derived from a Risk-Management Approach to Climate Change", *Climatic Change,* 95: 325-339, 2009.
Ackerman et al. (2009) criticize optimization applications of integrated assessment models of climate change on several grounds. First, they focus attention on contestable assumptions about the appropriate discount rate. Second, they worry that integrated assessment models base their damage estimates on incomplete

information including questionable estimates of the value of human life and/or ecosystem services. Thirdly, they suggest that mitigation costs are systematically over estimated because they ignore technological innovation.

So what good is economics in the climate arena? The authors suggest only one opportunity—investigate how the cost of achieving politically or hedging-based climate targets might be minimized. Their contribution provides a concise and internally consistent presentation of several sources of concern, but none is really new in their fundamental arguments. Antecedents of the points that they raise (and many others, for that matter) can be found in the established literature from the past 5 or 10 years; see, for example, Yohe (2003), Weitzman (2009) or Yohe and Tol (2008).

**#152** Kunreuther, H, Heal, G., Allen, M., Edenhofer, O., Field, C., and Yohe, G., "Risk Management and Climate Change", *Nature Climate Change,* March 2013.

The selection of climate policies should be an exercise in risk management reflecting the many relevant sources of uncertainty. Studies of climate change and its impacts rarely yield consensus on the distribution of exposure, vulnerability or possible outcomes. Hence policy analysis cannot effectively evaluate alternatives using standard approaches, such as expected utility theory and benefit-cost analysis. This Perspective highlights the value of robust decision-making tools designed for situations such as evaluating climate policies, where consensus on probability distributions is not available and stakeholders differ in their degree of risk tolerance. A broader risk-management approach enables a range of possible outcomes to be examined, as well as the uncertainty surrounding their likelihoods.

# CHAPTER 10

# MAKING ITERATIVE RISK MANAGEMENT WORKABLE IN A CHANGING CLIMATE

*The lede: it is very difficult, if not impossible, accurately to estimate the likelihoods of some events associated with climate change. Moreover, it is sometimes equally difficult to calibrate their consequences. Still, looking through even a qualitative, risk-based lens can be enormously useful for making, implementing, evaluating and/or interpreting policy even in the most challenging of circumstances. This means considering the elements of risk - consequences and likelihoods - one at a time, and then together.*[9]

The IPCC provided some guidance to its authors about handling the challenges of consistently accounting for uncertainties with regard to specific findings. The guidance proposes organizing one's thoughts around two distinct criteria:

> First, confidence in a particular finding that can be assessed by evaluating the "type, quality and consistency" of the available evidence and data from which the finding or hypothesis was drawn.

> Secondly, it is also important to assess the "degree of agreement" in understanding what is driving a finding and describing rigorously the underlying causal processes behind those understandings.

To take one example, there are lots of quality data demonstrating that gravity works; and there is widespread acceptance of the physics behind why. So, if you climb the Leaning Tower of Pisa and drop an apple, you can believe that somebody (maybe not you) can accurately project how long it will take for it to hit the ground and where it will land. Maybe they can even project whether or not it will bounce.

---

[9] Much of this chapter is derived directly (and sometimes word for word) from my Module 5, prepared for the Climate Judicial Project, inspired by Justice Stephen Breyer before he stepped down from the Supreme Court.

To take another more substantive example, macroeconomic data are collected across the United States every day, week, or month. They are extremely well respected. Moreover, the distributions that characterize their behavior for macro scale metrics, like inflation and unemployment, are widely understood and accepted. They can even be displayed as charts by Stephen Rattner on Morning Joe (MSNBC).

Nonetheless, there are two conflicting perspectives about how the economy works. One is "neo-Keynesian" (wherein federal spending can sustain employment). The other is "monetarist" (wherein federal spending can only generate inflation). When you ask whether the post-COVID outbreak federal spending under the American Rescue Plan was a good idea, monetarists will say "no!" and neo-Keynesians will say "for sure!"[10] Two years later, inflation was at a 50 year high and unemployment was at a 75 year low. Six months later, inflation was falling after the Federal Reserve Bank of the US had reduced the money supply and increased interest rates. So, either explanation could be right, but certainly cannot be completely achieved.

Thus, we have the roots of a quandary. How can we assess confidence in a particular scientific or economic findings consistently and comparably? Figure 10-1 shows the IPCC approach by plotting limited, medium, or robust evidence against low, medium, or high agreement to establish the foundations for five different levels of confidence: very high, high, medium, low, and very low. These are judged relative to the scale on right hand side of the figure.

Exactly where the divisions lie is, of course, subjective. Within this matrix, movement toward higher agreement but lower levels of evidence can increase or decrease confidence in a finding depending upon the relative subjective weights placed on agreement and evidence. Moving toward lower agreement with greater evidence can similarly move confidence in either direction, but moving simultaneously toward higher (lower) agreement and stronger (weaker) evidence means higher (lower) confidence. For our macroeconomic example, my take is little more than medium confidence in the lower right-hand corner.

Why should we worry at all about findings with "low", "very low", or even "medium" confidence assessments? Some climate-related events have a low likelihood but could generate enormous impacts (a runaway greenhouse

---

[10] https://www.whitehouse.gov/american-rescue-plan

effect from melting tundra that releases a greenhouse gas called 'methane'. Or the collapse of large portions of the West Antarctic Ice Sheet with associated sudden and large sea level rise.

Considering these events, despite their "low" or "very low" confidence assessments, enables decision-makers and opinion-makers to focus attention on them in accordance with what iterative risk management requires. Consider, as an example, the possibility of a runaway greenhouse effect caused by accelerated out-gassing of methane or carbon dioxide from melting tundra in the Arctic that would have enormous consequences across a wide range of measures. This would not be something to ignore; and risk management means doing so is not fear mongering.

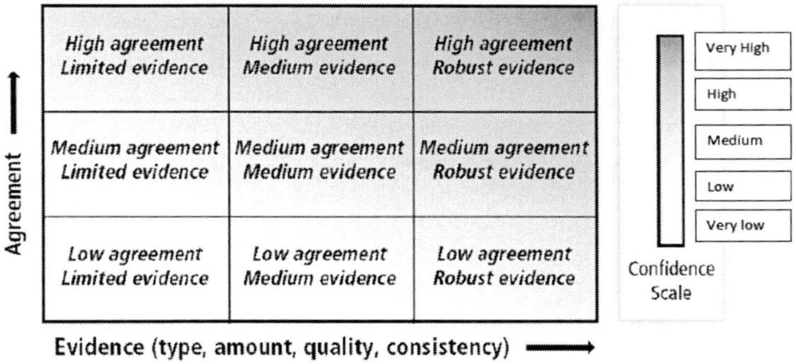

**Figure 10-1. Confidence assessments depend on the quality of the underlying evidence and on agreement in understanding underlying processes that can be contingent on a wide range of possible futures** (for example, a trajectory limiting temperature increases to 1.5 or 2.0 or 3.0 degrees Centigrade). Source: http://gyohe.faculty.wesleyan.edu/files/2018/05/125.pdf .

Runaway greenhouse warming is just one of a small collection of "large-scale singularities" that the countries of the world would view as "dangerous interference in the climate system", per Article 2 of the United Nations Framework Convention on Climate Change.[11] They are all part of the fifth Reason for Concern from chapter 6, and they are part of the reason why iterative risk management achieved consensus, per chapter 8.

It is interesting to note, in passing, that this approach to low-confidence/high-impact events is very similar to and uses the same language as the U.S. and

---

[11] https://unfccc.int/resource/docs/convkp/conveng.pdf

United Nations intelligence communities in assessing and reporting its findings.[12] There, analysts assess the credibility of the source of a finding given corroboration or inconsistencies in the underlying explanations. They also assess the quality of the underlying information, paying careful attention to whether it is direct evidence or circumstantial. Perhaps placing the summary information in both dimensions in a matrix like Figure 10-1, they can offer confidence assessments from very high to very low – and they report very low confidence findings to the highest levels of government and homeland security if the consequences could be very high.[13]

## The use of new knowledge in iterative risk management

The concepts of confidence and iterative risk management can be especially helpful when thinking about incorporating new science into an existing body of knowledge. "Iterative" is the critical adjective here. If a finding were dominated by temporal uncertainty, then would be it prudent to plan to make mid-course corrections based on new information about what is going on in the world. More specifically, the critical issue is when and how to incorporate new science into conventional wisdom so that it may or may not advance, reinforce, or sometimes provide reasons to doubt decades of accepted practice.

To cite a specific case, return to the example mentioned earlier: a runaway greenhouse effect caused by accelerated out-gassing of methane or carbon dioxide from melting tundra in the Arctic. What can be said about the confidence of such an event occurring? What would happen to that confidence if new science seemed to undercut an earlier consensus? How should such news be communicated?

In 2007, the IPCC reported high agreement from land surface models that the extent of Arctic permafrost will decline during the 21st century, in large measure because of particularly rapid warming across the region. In 2014, the next IPCC assessment continued to voice concern about a positive feedback loop — that melting permafrost could release increasing amount of greenhouse gases into the atmosphere and thereby accelerate the pace of the future warming. In 2020, though, a Canadian author team concluded from newly analyzed paleoclimatic data that the frozen store of carbon

---

[12] https://www.dni.gov/files/documents/ICD/ICD%20203%20Analytic%20 Standards.pdf

[13] https://hr.un.org/page/conclusions-and-preparing-investigation-report

release might not be so sizable.[14] In accordance with the IPCC confidence language protocol, this finding reduced confidence in the conclusion that the planet may experience a runaway methane emissions scenario.

It also brought another question to the fore: how should new science about the possible release of methane released from permafrost be communicated to decision-makers and influence-makers who have adopted a risk management approach? The answer to this question should begin with the observation that new science rarely contradicts current understanding completely. The conclusion of one new study being wildly different from conventional wisdom does not imply the value of giving extra weight in the first instance. In the vast majority of cases, it simply means altering assessors' subjective views about the confidence with which a former conclusion is held — both with respect to likelihood and consequence.

In this case, views of the consequences of a runaway greenhouse effect were not altered. Rather, the likelihood that future methane releases and associated damages would be small was increased. As a result, the assessment team could conclude that a more benign future was now a little more likely. However, they could add that runaway warming from permafrost melting was still possible, and that continued research into the matter would be prudent.

The news is not always this good. Recent science from Antarctica suggests two very troubling findings, wherein it could be much worse. First, the pace of the contribution of routine melting from the West Antarctic Ice Sheet has increased with statistical significance over the past decade or so. New sea level rise estimates reflect that finding in their upper scenario with high confidence, associated with an extra increase in global sea rise of about one foot through 2100. Secondly, more limited time-series observations of the ice structures that support the Thwaites glacier strongly suggest that those structures are disintegrating quickly, and perhaps irreversibly. That potential raises the possibility of a second extra increase in sea level rise of more than a foot, which could appear around the world within a decade or so, starting five years ago.[15] The first finding is troubling; the second could be catastrophic, because the timing could be so wildly accelerated.

---

[14] https://www.doi.10.1126/science.aax0504 .

[15] https://yaleclimateconnections.org/2021/04/with-seas-rising-stalled-research-budgets-must-also-rise/

## Using risk matrices

To accommodate situations like the last example, the context of the potential costs of climate change has come to reinforce the conventional definition that risk is the product of the likelihood of an uncertain event actually occurring multiplied by the consequences of that event. In response the critical need to reflect this definition, risk matrices have emerged as one of the most effective tools by which analysts, scientists, decision-makers, opinion-makers, judges, and others can communicate clearly with one another.[16] Figure 10-2 shows how a matrix can be used to project current and projected future risks in terms of likelihood and consequence.

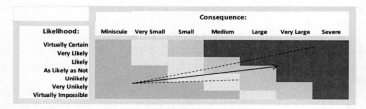

**Figure 10-2. The conceptualization of risk as the product of likelihood and consequence can be portrayed by a two-dimensional matrix.**

Lightly colored boxes in the lower left bottom corner identify low-risk combinations of likelihood and consequence; they are relatively benign and unlikely, so they are of relatively small concern. Slightly lighter boxes suggest moderate concern, while the slightly darker boxes north and to the right capture combinations that fall just short of the darkest combinations of major concern.

Figure 5-2 also suggested how this conceptual device could be used to account for uncertainty about the future in the depiction. The upper dotted line represents a hypothetical 95th percentile scenario that portends larger consequences with growing likelihood sooner than the baseline. The lower dotted line represents a 5th percentile scenario trajectory. It is shorter, because it tracks below the median and thereby depicts cases where the consequences of climate change increase more slowly.

---

## Attitudes toward tolerable risk

The term "risk" has many definitions in colloquial use, and one is particularly useful - consider "*tolerable risk*". It can be seen as the level of risk deemed acceptable by a society or by an individual in order to sustain some particular benefit or some critical level of functionality. Achieving tolerable risk does not mean eliminating all chance that bad events will occur. It does not even mean that the damage from an event, such a storm of wildfire, will not be catastrophic to some people. It does, however, mean that risk has been evaluated and is being managed to a level of comfort acceptable according to the particular risk aversion applied.

Public health provides another example from the United States. Figure 10-3 shows mortality proportions by state from ordinary flu.[17] The categories illustrated by color indicate different levels of tolerable risk with which citizens have become comfortable. How do we know? Because those citizens are not demanding more, even though the means are available.

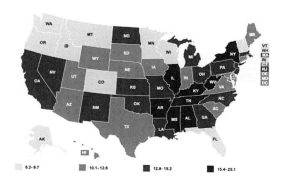

**Figure 10-3. Reported deaths from Asian flu per 100,000 people across 50 states.**

The New York (City) Panel on Climate Change (NPCC 2010) employed this notion to frame both its evaluation and its management of climate

---

[17] https://www.kff.org/other/state-indicator/influenza-and-pneumonia-death-rate/?activeTab=map&currentTimeframe=0&selectedDistributions=influenza-and-pneumonia-deaths&sortModel=%7B%22colId%22:%22Location%22,%22sort%22:%22asc%22%7D

change risks to public and private infrastructure.[18] NPCC communicated this concept to planners and decision-makers across the city, by pointing out that building codes imposed across the city did not try to guarantee that a building would never fall down. Instead, they were designed to produce an environment in which the likelihood of a collapse was acceptable, given the cost and feasibility of doing more to prevent collapse. However, if climate change or another stressor pushed a particular risk profile closer to the thresholds of social tolerability by increasing likelihoods, investments in risk-reducing adaptations would be expected to increase.

For many reasons, achieving broad acceptance for any tolerable risk threshold across a population is a huge task. For one, risk tolerance varies widely across societies and individuals. For another, policymakers confronting a pandemic or extreme climate change are often navigating between different, and perhaps strongly contradictory, risk management priorities.

Such challenges can be met by individuals and communities and institutions agreeing on a common set of facts about likelihood and possible consequences of an event. They can then explore together how much risk is acceptable and why. During the second wave of the COVID-19 pandemic, for example, New York State relied on science to frame its strategies to avoid overwhelming its hospital system after what had been a relatively successful initial response. It implemented two forward-looking criteria: (1) hold the transmission rate of the virus below 1.0 and (2) keep vacancies of hospital beds and ICU beds across the state above 30 percent of total bed capacity. These represented thresholds of tolerable risk that could be monitored and projected into the future, using results from integrated epidemiological-economic models.

## Insuring Against Risks

One way of dealing with risks is through insurance. Sometimes, insurance comes in the form of hedging against a bad event by being careful to lower the likelihood of an event. Other times, it comes in the form of taking actions to lower the consequences of an event. In either case, insurance is a form of investment in which money spent now yields future benefits - but insurance is not a panacea.

Consider, for example, coastal residents' insurance-based responses to increased flooding risks caused by human-induced sea level rise. Suppose

---

[18] https://www.nyas.org/annals/climate-change-adaptation-in-new-york-city/.

that the residents (the demand side of a flood insurance market) think that the premium offered by the insurer and certified by a regulator is too high — that is, they think that the quoted premium is greater than any credible estimate of the probability of loss from flooding.

Perhaps their subjective confidence in climate science or in the connection between impacts and socioeconomic consequences is low. Or, perhaps they have grown accustomed to premiums that were calculated on the basis of historical losses and not on the basis of projected future losses — losses that science now projects to become larger and more frequent in the future. They may also think that they are entitled to the subsidized system in which the Federal Emergency Management Agency (FEMA) covers any and all extreme damages at a minimal participation fee under the National Flood Insurance Program (NFIP).[19]

For any or all of these reasons, a resident could underinsure, meaning that those residents would not be paying at a rate that will be sufficient to cover actual damage. This leaves damage to repair and no source of funds to pay for it. The question arises of who should pick up the cost. Since, in conventional economics terms, rational individuals would take advantage of properly priced insurance opportunities in their own best interest, society should not pay for losses that could have been covered by individuals. This general observation, applicable beyond the specific context of insurance, can be applied in the climate change context.

Of course, insurance companies need information about the likelihoods and consequences of climate-related events if they are to write policies to cover potential damages. In some cases, insurance coverage is not feasible because the risks are not known or, if they are known, have not made public by the experts who have investigated them. One more significant problem is becoming widespread: the past is no longer the prologue of the future.

Recent experiences in the United States have underscored this point. The 2020 wildfires in California, the hurricanes of 2020, Hurricane Ian in 2022, and the combination of atmospheric rivers and bomb-cyclones across mors of the United States all illustrate how extreme events are increasing in intensity and frequency; this is true especially when it is recognized that

---

[19] https://corporate.findlaw.com/law-library/how-the-emergency-national-flood-insurance-program-works.html?DCMP=google:ppc:K- FLPortal:10313486553: 103002902536&HBX_PK=&sid=1014715&source=google~ppc

Europe was, at the *very same time,* experiencing record heat for weeks at a time.

Coincidence is not a laughing matter. Climate impacts pile on and happen simultaneously. In California, for example, of the state's largest 20 fires (in acres burned), only three had occurred prior to 2000; nine of the biggest 10 occurred in the nine years since 2012. In 2017, 9,270 fires burned a record 1.5 million acres. The Mendocino Complex fire in 2018 became the (then) largest wildfire in California's history. Historic drought and unprecedented heat marked 2022, even while never-before-seen rain and associated flooding occurred elsewhere, or at the same place at a different time.[20]

The decade finished with a less noteworthy year in 2019, but then came 2020. A new largest fire in California history, the Complex fire, started in August 2020. Soon thereafter came the third, fourth, fifth, and sixth largest wildfires in the state's history. By October 3, these five conflagrations and nearly 8,000 other more "ordinary" wildfires had killed 31 people and burned more than four million acres. Incredibly, on that day, all five of those fires were still burning.[21]

Human actions are, of course, a major factor of increased fire risk. On the consequences side of the risk calculation, catastrophic damage to life and property has increased markedly as more people have moved into vulnerable forested areas, putting their lives and property at risk and setting more inadvertent blazes. Changes in forest management have also contributed, because fire suppression policies reduced the frequency of blazes that could reduce the fuel reserves built up in forests. However, these non-climate contributors to increased fire danger have not increased sufficiently to fully account for the recent devastation.

The change in the various individual factors that create wildfire threats cannot explain the devastation if taken one at a time. Many of the 2020 fires were caused by a record number of dry lightning strikes. They were not solely the result of climate change, but they fed into a witches' brew of conditions that are all linked to global warming. The lightning strikes and other points of ignition hit in the midst of a record drought and heat wave that had lasted for weeks on end. Years of bark-beetle infestations had

---

[20] https://yaleclimateconnections.org/2021/09/never-before-nb4-extreme-weather-events-and-near-misses/
[21] CalFire (2021) Incidents and events – the 20 largest California wildfires, https://www.fire.ca.gov/media/4jandlhh/top20_acres.pdf

produced large stands of dead trees, because warmer winters had increased springtime beetle populations. Also, decades of gradual warming had extended the western fire season by some 75 days. Taken together, these contemporaneous influences reveal that the issue is not just what sparks the fires. The larger problem is the context in which they started, and how quickly they spread once started.

A similar story can be told about damage from tropical storms. Hurricanes Harvey in 2017 and Florence in 2018 dropped historic amounts of rain after making landfall and then stalling over Houston and North Carolina respectively. In the summer of 2020, hurricanes Laura and Beta followed suit, causing extreme rainfall totals and substantial damage from storm surge. Their behaviors mimicked Dorian over the Bahamas in 2019. Finally, in 2017, Hurricane Mike traveled more than 100 miles inland from landfall along the Florida panhandle, only to stall over Albany, Georgia long enough to deposit in excess of 5 feet of rain in some locations, and around 4 feet across half of the state.[22]

Near-record high ocean and gulf temperatures have allowed more tropical depressions and non-tropical low-pressure systems to develop into dangerous hurricanes. The water temperature in the Gulf of Mexico when Hurricane Ian was intensifying off of the western coast of Florida was above 90 degrees Fahrenheit. At the same time, the decrease in the summer temperature difference between the Arctic and the tropics has weakened steering currents in the atmosphere, causing storms to move more slowly. In addition, sea-level rise, one of the most obvious results of decades of rising temperatures, has compounded risks posed by storm surge.

The expanding consequences of compound fire and flood events are also having negative effects. Many of the worst fires and hurricanes have exploded so quickly and have spread so erratically that human evacuations have become "moment's notice" emergencies. As with residents of the southeastern and Gulf coasts, residents from California and Oregon have had to retreat from harm's way as quickly as possible.

New information from the ongoing scientific process can open new doors of inquiry, to be sure. More usually, though, as noted above, new evidence hardly ever implies either dismissing conventional wisdom entirely or reversing its content. In any case, it is critical that some protocol like what

---

[22] https://yaleclimateconnections.org/2020/10/multiple-extreme-climate-events-can-combine-to-produce-catastrophic-damages

has become standard in IPCC assessments be followed in bringing the new science into existing assessments. Only then will new reports affecting the confidence with which a piece of conventional wisdom is held have credibility.

# CHAPTER 11

# SNOWMASS INTEGRATED ASSESSMENT MEETINGS

*The lede: starting in 1995, John Weyant, James Sweeney, Alan Manne, Richard Richels and other friends organized and secured funding for two weeks of annual meetings on the integrated assessment of climate impacts. The meetings were convened every August for nearly two decades in Snowmass, CO – at at about 9500 feet of altitude, held under a tent. $1500 winter skiing condos went for less than $200 in August. That was a good deal and the venue was wonderful. Summer wildflowers blooming on winter ski slopes intermingled with Aspen trees are just marvelous.*

Funding for the meetings came from an eclectic mix of public and private money: the National Science Foundation, Japan, Chevron, Exon-Mobile, the Electric Power Research Institute, and many other sources. Nobody got paid, but housing expenses and meals were covered. The rules were clear - one-third of the attendees each year would be new participants from new academic fields (not the usual suspects) so that discussions about integrated assessment, impacts, and climate risks would always see new perspectives. Some participants never returned from their first times under the tent, but others engaged in discussions and returned year after year. Participants presented their work, and then we all collaborated on new science.

From 1995 through 2016 when the meetings were studied, more than 1000 published peer-reviewed articles had been produced from the interactions and collaborations that were consummated in Snowmass. I can trace more than 20 papers on my CV to collaborations that began in Snowmass, including some with many more than 250 publications.

As part of the organizing efforts, I worked with Alan Manne and others to create and sustain the "Uncertainty Working Group" for many years. Our task was to get integrated assessment modelers to agree to run specific input trajectories for what we all agreed were critical drivers of climate change and climate impacts. Their models would produce their result given those trajectories, and so we could compare models. Of course, modelers could also report alternative results designated as "modelers' choices" base on different drivers.

We discovered that variances in outcomes (emissions, temperature, etc...) were larger across the participating modelers when they accepted common drivers than it was when they could choose their own inputs. Apparently, modelers were truncating drivers' distributions so that their results would not be outliers in the grand scheme of things. This was big news, and lead many modelers to change their research protocols.

In addition to the organizers, my memory of collaborators that were attracted to Snowmass include Susan Sweeney, Richard Tol, Jae Edmonds, Hugh Pitcher, Terry Rood, Stephen Schneider, Natasha Andronova, Michael Schlesinger, Brian O'Neill, William Nordhaus, Kenneth Strzepek, Richard Moss, Jerry Melillo, Jake Jacoby, Sally Kane, Kritie Ebi, Thomas Willbanks, Anthony Janetos, Saleemul Huq, Linda Mearns, Hadi Dowlatabadi, Camille Parmesan, Joel Smith, William Easterling, and many more.

Snowmass was also family time. My younger daughter Courtney worked for Susan Sweeney to make several of the two-week sessions work well. She also learned to ride a horse from Alan Manne. I played golf with many at altitude – what a hoot. The Snowmass course had a tee shot nearly 100 feet above a fairway on the back nine. A good tee shot would fly up in the ski against a background of the mountains and hang in the air for what seemed like many minutes. More importantly to the golfer in me, it would not roll into the guarding lake, because it landed going straight down.

I also had a standing $1 bet with Jim Sweeney on women's basketball games between UCONN and Stanford; we also agreed to pay $1 if UCONN or Stanford won a national championship, even if they had not played each other during the season. Except for the Tampa Final Four game, I collected my winnings publicly under the tent at the beginning of one of my annual presentations. One dollar a year piled up year after year, so Jim and I negotiated a present value calculation to end the embarrassment. Jim paid me twenty dollars (with a negotiated 5% discount rate), and that ended my tendency to show a picture of UCONN cutting down the nets as the first slide in my presentations.

Many of the new participants were young scholars. I always sat at the back of the tent, and so did they. At a break, I frequently spoke with one or more who conveyed the impression of most of the 'newbies': "half of the bibliography of my dissertation is sitting under this tent. What do I do?" I would respond, "introduce yourself and see what happens!". Many would hold out their hand and say "I'm…. and I have read your stuff." That would be my cue to have a chat and then introduce them to the others. They went on to contribute many of our 1000 papers.

# References, abstracts and links connected to my January 2023 CV

**#52** Strzepek, K., Yates, D., Yohe, G., Tol, R. and *Mader, N., "Constructing "Not Implausible" Climate and Economic Scenarios for Egypt", *Integrated Assessment* 2: 139-157, 2001.

A space of "not-implausible" scenarios for Egypt's future under climate change is defined along two dimensions. One depicts representative climate change and climate variability scenarios that span the realm of possibility. Some would not be very threatening. Others portend dramatic reductions in average flows into Lake Nassar and associated increases in the likelihood of year to year shortfalls below critical coping thresholds; these would be extremely troublesome, especially if they were cast in the context of increased political instability across the entire Nile Basin. Still others depict futures along which relatively routine and relatively inexpensive adaptation might be anticipated. The ability to adapt to change and to cope with more severe extremes would, however, be linked inexorably to the second set of socio–political–economic scenarios. The second dimension, defined as "anthropogenic" social/economic/political scenarios describe the holistic environment within which the determinants of adaptive capacity for water management, agriculture, and coastal zone management must be assessed.

**#57** Yohe, G. and Tol, R., "Indicators for Social and Economic Coping Capacity – Moving Toward a Working Definition of Adaptive Capacity", *Global Environmental Change* 12: 25-40, 2002.

This paper offers a practically motivated method for evaluating systems' abilities to handle external stress. The method is designed to assess the potential contributions of various adaptation options to improving systems' coping capacities by focusing attention directly on the underlying determinants of adaptive capacity. The method should be sufficiently flexible to accommodate diverse applications whose contexts are location specific and path dependent without imposing the straightjacket constraints of a ''one size fits all'' cookbook approach. Nonetheless, the method should produce unitless indicators that can be employed to judge the relative vulnerabilities of diverse systems to multiple stresses and to their potential interactions. An artificial application is employed to describe the development of the method and to illustrate how it might be applied. Some empirical evidence is offered to underscore the significance of the determinants of adaptive capacity in determining vulnerability; these are the determinants upon which the method is constructed. The method is, finally, applied directly to expert judgments of six different adaptations that could reduce vulnerability in the Netherlands to increased flooding along the Rhine River.

**#58** Yohe, G. and Schlesinger, M., "The Economic Geography of the Impacts of Climate Change", *Journal of Economic Geography* 2: 311-341, 2002.
Our ability to understand the geographical dispersion of the impacts of climate change has not yet progressed to the point of being able to quantify costs and benefits that are distributed across globe along one or more climate scenarios in any meaningful way. We respond to this chaotic state of affairs by offering a brief introduction to the potential impacts of a changing climate along five geographically dispersed portraits of how the future climate might evolve and by presenting a modern approach to contemplating vulnerability to climate impacts that has been designed explicitly to reflect geographic diversity. Three case studies are offered to provide direct evidence of the potential value of adaptation in reducing the cost of climate impacts, the versatility of thinking about the determinants of adaptive capacity for specific regions or sectors, and the feasibility of exploring both across a wide range of 'not-implausible 'climate and socio-economic scenarios. Three overarching themes emerge: adaptation matters, geographic diversity is critical, and enormous uncertainty must be recognized and accommodated.

**#70** Schlesinger, M., Yohe, G., Yin, J., Andronova, N., Malyshev, and Li, B., "Assessing the Risk of a Collapse of the Atlantic Thermohaline Circulation" in *Avoiding Dangerous Climate Change* (Schellnhuber, H.J., Cramer, W. Nakicenovic, N. Wigley, T., and Yohe, G. eds.), Cambridge: Cambridge University Press, 2006.
In this paper we summarize work performed by the Climate Research Group within the Department of Atmospheric Sciences at the University of Illinois at Urbana-Champaign (UIUC) and colleagues on simulating and understanding the Atlantic thermohaline circulation (ATHC). We have used our uncoupled ocean general circulation model (OGCM) and our coupled atmosphere-ocean general circulation model (AOGCM) to simulate the present-day ATHC and how it would behave in response to the addition of freshwater to the North Atlantic Ocean. We have found that the ATHC shuts down 'irreversibly' in the uncoupled OGCM but 'reversibly' in the coupled AOGCM. This different behavior of the ATHC results from different feedback processes operating in the uncoupled OGCM and AOGCM. We have represented this wide range of behavior of the ATHC with an extended, but somewhat simplified, version of the original model that gave rise to the concern about the ATHC shutdown. We have used this simple model of the ATHC together with the DICE-99 integrated assessment model to estimate the likelihood of an ATHC shutdown between now and 2205, both without and with the policy intervention of a carbon tax on fossil fuels. For specific subjective distributions of three critical variables in the simple model, we find that there is a greater than 50% likelihood of an ATHC collapse, absent any climate policy. This likelihood can be reduced by the policy intervention, but it still exceeds25% even with maximal policy intervention. It would therefore seem that the risk of an ATHC collapse is unacceptably large and that measures over and above the policy intervention of a carbon tax should be given serious consideration.

**#74** Tol, R. and Yohe, G., "A Review of the *Stern Review*", *World Economy* 7: 233-250, 2006.

The Stern Review on the Economics of Climate Change (Stern et al., 2006) was delivered to the Prime Minister and the Chancellor of the Exchequer of the United Kingdom in late October of 2006. A team of 23 people, led by Sir Nicholas Stern and supported by many consultants, worked for a little over a year to produce a report of some 575 pages on the economics of climate change, and their work has certainly drawn substantial attention in the media. Across its 575 pages, the Stern Review says many things, and some of the points are supported more strongly and developed more completely than others. Naturally, we agree with some of its conclusions, including the fundamental insight that there is an economic case for climate policy now, and that the cost of any climate policy increases with delay; this is, of course, not really news.

We do, though, disagree with some other points raised in the Stern Review; here, we raise six issues. First, the Stern Review does not present new estimates of either the impacts of climate change or the costs of greenhouse gas emission reduction. Rather, the Stern Review reviews existing material. It is therefore surprising that the Stern Review produced numbers that are so far outside the range of the previous published literature.

Second, the high valuation of climate change impacts reported in the Review can be explained by a very low discount rate, risk that is double-counted, and vulnerability that is assumed to be constant over very long periods of time (two or more centuries, to be exact).

Third, the low estimates for the cost of climate change policy can be explained by the Review's truncating time horizon over which they are calculated, omitting the economic repercussions of dearer energy, and ignoring the capital invested in the energy sector.

Fourth, the cost and benefit estimates reported in the Stern Review do not match its policy conclusions

Fifth, a strong case for emission reduction even in the near term can nonetheless be made without relying on suspect valuations and inappropriate summing across the multiple sources of climate risk. A corollary of this observation is that doing nothing in the short term is not advisable even on economic grounds.

Sixth, alarmism supported by dubious economics born of the Stern Review may further polarize the climate policy debate.

**#84** Yohe, G., Tol, R. and Murphy, D., "On Setting Near-term Climate Policy while the Dust Begins to Settle: The Legacy of the *Stern Review*", *Energy and Environment* 18: 621-633, 2007.

We review the explosion of commentary that has followed the release of the *Stern Review: The Economics of Climate Change*, and agree with most of what has been written. The Review is right when it argues on economic grounds for immediate intervention to reduce emissions of greenhouse gases, but we feel that it is right for the wrong reasons. A persuasive case can be made that climate risks are real and increasingly threatening. If follows that some sort of policy will be required, and the least cost approach necessarily involves starting now. Since policy implemented in 2007 will not "solve" the climate problem, near term interventions can be designed to begin the process by working to avoid locking in high carbon investments and providing adequate incentives for carbon sequestration. We argue that both objectives can be achieved without undue economic harm in the near term by pricing carbon at something on the order of \$15 per ton as long as it is understood that the price will increase persistently and predictably at something like the rate of interest; and we express support for a tax alternative to the usual cap-and-trade approach.

**#112** Keller, K., Tol, R.S.J., Toth, F.L., and Yohe, G., "Abrupt Climate Change near the Poles", *Climatic Change* 91: 1-4, 2008.
Natura non facit saltus—but with anthropogenic climate change it just might. This isone of the major reasons for concern about greenhouse gas emissions. The North Atlantic Meridional Overturning Circulation (NAMOC)—widely but incorrectly known as the thermohaline circulation—and the West-Antarctic Ice Sheet (WAIS) are both poster children of this concern about extreme climate change scenarios, signifying the potential for regional cooling and rapid global sea level rise. Yet, both scenarios are uncertain in the Knightian sense. We know that the NAMOC has slowed down or even collapsed in the past and that the WAIS can disintegrate. We do not know, however, how likely these scenarios are, let alone how emission abatement would affect that probability. We know even less about the consequences, only that they may be serious. Uncertainty is, of course, no reason for inaction—indeed, uncertainty can substantially strengthen the case for emission reduction. However, uncertainty also opens possibilities for confusion, scaremongering, or denial.

# CHAPTER 12

# THE SOCIAL COST OF CARBON
# (AND METHANE AND NITROUS OXIDES)

*The lede: the social cost of carbon (SCC) is an estimate of the damage caused by a ton of carbon emissions along a specific future emissions projection based on specifications of attitudes toward risk, discounting the future, climate sensitivity, global mitigation patterns, and so on. I was involved in helping the EPA to understand what it was, and what it was not, consulting with Stephen Rose and Benjamin Deangelis, for example. It was both a difficult and simple concept to accept. It was, though, the foundation of the Supreme Court's ultimate decision to classify carbon dioxide as a pollutant so that (1) the Clean Air Act applied to carbon emissions and (2) the social value of reduced carbon emissions from, for example, increased mileage standards for vehicle fleets should include legitimately calculated measures of the SCC.*

Current common practices generally use one published aggregate estimate of this economic damage (from a list of four alternatives) to anchor their approach to the second task (though they usually use the other three to set context). They would use "the social cost of carbon dioxide" (SC-$CO_2$), but they would call the social cost of carbon (thus its "SCC" notation). This estimate is defined as the value of the damage resulting from the emission of one more ton of carbon dioxide. It also can be interpreted as the damages that can be avoided by *removing* one ton of carbon dioxide from the emission stream (that is, focusing on the benefits rather than the costs). The SCC is the most widely used way to summarize aggregate estimates of the economic consequences from climate change impacts.[23]

---

[23] The difference between the SC-$CO_2$ values and the SCC is easily framed. SCC is equal to 3.67 times the equivalent value of the SC-$CO_2$ (because the ratio of the atomic weight of $CO_2$ to the atomic weight of carbon is equal to 44/12 or 3.67). Care must be taken to recognize the units when any calculation is undertaken, since the error of misreporting can be a multiple of almost 4.

Estimates of the social cost of methane ($SC\text{-}CH_4$) and the social cost of nitrous oxides ($SC\text{-}NO_x$) are analogous to the SCC, since they are also greenhouse gases that contribute to warming along different timescales.

The SCC was never intended to be an estimate of the efficient price of carbon; that calculation would require some characterization of the cost of mitigation. Ranges of estimates of the SCC are, though, appropriate for quantifying ranges of value added from a climate perspective of policies or programs that would reduce or increase carbon emissions as a side effect – for example, increased mileage standards (CAFÉ standards) for automobiles that would increase mileage and therefore reduce emissions per unit of service provided of lots of things. Some measure of the economic value of such an effect on carbon emissions should certainly be included in a benefit-cost calculation for such a policy or program proposal.

There came a day when discussions of the SCC were elevated to the highest level of the United States executive branch. Lawrence (Larry) Summers, then the Secretary of the Treasury for President Obama who regularly attended the daily 6:30 AM presidential briefings, had seen some estimates that reported a negative SCC (that is, a benefit from warming). "If that is true, why are we worried about this?" was his question at a morning briefing with the President. Indeed, 12% of published estimates were negative – because of very high discounting and what turned out to be exaggerated assumptions about increased farm production caused by "$CO_2$" fertilization (plants eat carbon dioxide for photosynthesis, so higher concentrations could make farm crops more productive).

The other 88% of estimates were positive, and some were significantly so. The call went out to respond to his question by focusing on the distribution of SCC estimates (ranging from -$12 per ton of carbon emitted to +$300 per ton and more). Figure 11-1 replicates the 2021 version of the graph that the Secretary had seen. The question arose: why was a negative SCC estimate possible (agriculture and energy benefits in the short run with a very high discount rate because the long-term is bleak) but not plausible (because CO2 fertilization would peak, agriculture needs water, and high discount rates are inappropriate)? We were asked to provide a collection of papers that were on these points.

The President was known to take 300 pages of academic reading to the residence for bedtime reading in preparation of the next day's 6:30 AM briefing. I got such a call from Anthony Janetos within minutes of the

Secretary's intervention – drop everything and deal with this (please) and submit something before COB.

I had been important in developing the EPA's understanding of what the SCC meant and how it should be calculated and used. I remember that they interrupted a shopping trip to the Westbrook outlets; I was not popular at home at that point in time. Anyway, I wrote a concise paragraph for Tony, and sent along some papers for reading (numbers 92, 115, 118, and 128); and I taught the SCC in class that day as I way framing my response.

What did I teach in class that day because my preparation time was sent by the wayside? The definition of the social cost of carbon to both an intro class and an environmental clas with only hints of the story behind the interruption to their syllabus. The students never knew the details of the story behind why the syllabus had been changed on a dime, but it all worked out. I always taught from first principles, and SCC was a perfect opportunity.

The next day, our response to Secretary Summers was presented to the President at the morning briefing at 6:45. The Secretary was engaged, and wanted some reading. So, the papers that I sent were among the 300 pages of reading that President Obama took up to the residence that night. These were not 300 pages of press coverage and newspaper clippings; these were 300 pages of scholarly work published in the peer reviewed scientific literature. It was not a staffer doing the reading and providing a one-pager summary; it was the President of the United States.

Questions that I got after the briefing the next day (at 7:30 AM) indicated that the President and the Secretary had done their reading, synthesized its content, and engaged in a discussion.

You cannot imagine what it felt like to know that dropping everything really mattered. The President was pleased that Secretary Summers agreed that climate change was not a positive thing. His questions had been exactly on point; and his accepting science and economics was not a surprise. He never again objected to claims that climate change was a risk to humanity. And he has agreed with Janet Yellen that financial stress tests for major banks across North American should be driven by climate risks to coastal real estate obscured in bundles of mortgages – remember the debacle of 2009?[24]

---

[24] The US Federal Reserve Board (FED) raised this risk in Box 4 of its Nov 9, 2020 "Financial Stability Report" wherein they warned without notice about risk

Moving along, and based on a median estimate of SCC, CAFÉ standards for vehicle fleets were increased from 32 to 36 mpg, because the economic value of reduced carbon emissions made it economically worthwhile. The Department of Transportation was taken to court on this decision. The penultimate hearing was an appeals court in Massachusetts. Justice Sotomayor participated in a three-judge panel that decided that carbon dioxide was a pollutant. The automobile industry took that decision to the Supreme Court. Justice Sotomayor had been promoted by that time, so she recused herself. The Supreme Court, in a 5-3 decision for which Chief Justice Roberts wrote the decision, agreed that carbon dioxide fell under the Clean Air Act as a pollutant. Nobody, until Trump and Pruitt, objected.

Estimates of the SCC are obviously based on climate and economic models that incorporate representations of numerous natural, physical, social, and economic systems. Relevant assumptions in the creation of these models include normative and behavioral parameters such as personal or social consumption impatience, attitudes toward risk, and consumption sensitivities to income and relative prices. Estimates of the SCC are thus contingent on a long list of parameters and assumptions that must be understood.

This is the point of Figure 12-1. It shows specifically how estimates of any of these social costs are contingent on choices about discounting and risk aversion at the very least. They are also determined by other sets of assumptions from their underlying climate and economic modeling: policy interventions to reduce emissions (or not) and/or to ameliorate damages (or not), population growth profiles, underlying technological changes, international trading structures, the availability of economic resources, and so on.

---

of "abrupt repricing of assets" and the need to provide information to help markets reflect such risks in asset prices.

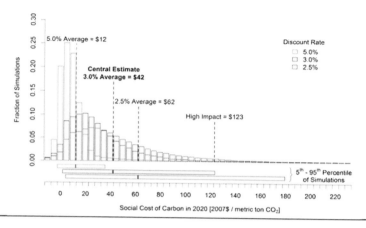

**Figure 12-1. Distributions of estimates of SC-CO₂ contingent on the discount rate with the justification of a high rate associated with high aversion to risk.**

Additionally, and importantly, social cost estimates also vary with time. Because the calculation is model-based, it is possible to make estimates for future dates. These estimates increase over time - not because any of the modeling sensitivities or parameter choices are assumed to have changed, rather, they grow because increasing damages projected into the future for any specific modeling configuration occur closer to the benchmark date of the estimate and its discounting function; and so they are discounted less robustly.

The time sensitivity of the estimates independent of discounting was displayed forcefully by the United States Environmental Protection Agency in 2016, when it reported mean estimates for the SCC (actually SC-CO₂) in five-year increments from 2015 through 2050 for three alternative discount rates — 2.5 percent, 3.0 percent, and 5.0 percent.[25] The average SC-CO₂ ranged from $11 to $56 in 2015, and from $26 to $95 in 2050.

Since the discount rate and geographic coverage were arbitrary choices, it was possible to manipulate social cost estimates for political gain. In 2020, for example, the General Accounting Office reported that the Environmental Protection Agency of the Trump Administration had begun to use discount rates of 3 percent and 7 percent in calculating SC-CO₂. EPA also had

---

[25] https://19january2017snapshot.epa.gov/sites/production/files/2016-12/documents/sc_co2_tsd_august_2016.pdf.

changed the calculation of the impacts to include only U.S. damages rather than considering the worldwide economic consequences.[26] Their new modeling produced dramatically SCC estimates: $6 for 2020 up to only $9 by 2050. Making these two discretionary changes therefore reduced the marginal social cost of carbon dioxide emissions to nearly zero — not because the economics required it, but because key discretionary assumptions were changed.

The Biden Administration worked, by Executive Order, to update those estimates on the basis of the lower discount rate and a return to global coverage.

Politics are not the only thing that changes over time. The economic and physical science environments of the modeling also evolve over time, and so social cost estimates need to keep up to stay relevant and robust. For example, a paper published in *Nature* in the late summer of 2022 reported in that "[o]ur preferred mean SC-CO$_2$ estimate is $185 per tonne of CO$_2$ ($44-413/t-CO$_2$: 5-95% range, 2020 US dollars) at a near-term risk-free discount rate of 2 percent. These values are 3.6-times higher than the then current US government's values of $51/t-CO$_2$".[27] The "so what?" and "why?" questions behind the import of this paper in a prestigious journal were covered with great care in the Washington Post (WAPO).[28]

The WAPO coverage is, in fact, a perfect place to start an assessment of how to incorporate this new information into an appropriate evaluation of what new results mean for decision-making; significantly different results in one new paper need to be considered, but they do not yet mean that conventional wisdom needs to be completely overturned – at least, not yet.

The point of this example is that science and social science evolves. The contribution of Working Group II to the IPCC's sixth assessment report in 2022 and the National Academies' review of the foundations of the social cost of carbon in 2017. Aldy et al. (2021) contributed a new but not so dramatically different analyses.[29] So, not much is changing? Not really. Climate related extreme events had exploded in the meantime.

The bottom line, therefore, is that distributions of the social cost of carbon,

---

[26] https://www.gao.gov/mobile/products/GAO-20-254.

[27] https://www.nature.com/articles/s41586-022-05224-9

[28] https://www.washingtonpost.com/climate-environment/2022/09-01/costs-climate-change-far-surpass-government-estimates-study says/

[29] https://www.science.org/doi/abs/10.1126/science.abi7813

methane, or nitrous oxides is just a snapshots in time. Useful, but not permanent, they are. Just like other metrics that support decisions, estimates iterate naturally and legitimately therefore usually grows over time. The economics is dynamic because the science is dynamic; and the science is dynamic because climate change is on steroids. None of this means that the government was lying 5 years ago; it was just working with the best information available at the time.

Despite these caveats, estimates of these social costs are the only vehicle for conveying aggregate damage information to deliberations of climate policy to decision makers across the U.S. and around the world. They have begun to play an important role in many climate related judicial proceedings. Cass Sunstein has, in fact, argued convincingly that keeping the distribution of values in line with modern science and economics is essential to legal durability of any related policy or judicial decision.[30]

Therefore, it is essential that we emphasize that social cost estimates complement risk management strategies, and that those strategies are designed to accommodate uncertainties. Using the social cost of carbon enables one to weigh the risks of emissions against the costs of mitigating them by setting, for example, a temperature target or an emissions target at a point in time. At a later time, the most attractive targets may have changed.

## References, abstracts and links connected to my January 2023 CV

**#92** Yohe, G., "Thoughts on 'The Social Cost of Carbon: Trends, Outliers and Catastrophes", 2007.
The updated meta-analysis of estimates of the social cost of carbon (SCC) offered by Tol (2007) is a welcome addition to the conversations about "What is new?" and "What should we do with so much uncertainty?" It offers five conclusions with diminishing persuasion; but perhaps its largest contribution is its differentiating list of the 211 estimates that can now be found in the literature. Simply compiling, segregating, and referencing this list is worthy of commendation for those who try to bring the social cost of carbon to bear on issues of climate policy.

**#115** Anthoff, D., Tol, R.S.J, and Yohe, G., "Risk Aversion, Time Preference, and the Social Cost of Carbon", *Environmental Research Letters* 4 (2–2): 1–7, 2009.
The Stern Review reported a social cost of carbon of over $300/tC, calling for ambitious climate policy. We here conduct a systematic sensitivity analysis of this result on two crucial parameters: the rate of pure time preference, and the rate of risk

---

[30] https://scholarship.law.upenn.edu/penn_law_review/

aversion. We show that the social cost of carbon lies anywhere in between 0 and $120,000/tC. However, if we restrict these two parameters to matching observed behavior, an expected social cost of carbon of $60/tC results. If we correct this estimate for income differences across the world, the social cost of carbon rises to over $200/tC.

**#118** Anthoff, D., Tol, R.S.J, and Yohe, G., "Discounting for Climate Change", *Economics: The Open-Access, Open-Assessment E-Journal*, 3: 2009-2024, 2009.

It is well-known that the discount rate is crucially important for estimating the social cost of carbon, a standard indicator for the seriousness of climate change and desirable level of climate policy. The Ramsey equation for the discount rate has three components: the pure rate of time preference, a measure of relative risk aversion, and the rate of growth of per capita consumption. The focus of much of the attention on the appropriate discount rate for long-term environmental problems has been on the role played by the pure rate of time preference in this formulation. We show that the other two elements are numerically just as important in considerations of anthropogenic climate change. The elasticity of the marginal utility with respect to consumption is particularly important because it assumes three roles: consumption smoothing over time, risk aversion, and inequity aversion. Given the large uncertainties about climate change and widely asymmetric impacts, the assumed rates of risk and inequity aversion can be expected to play significant roles. The consumption growth rate plays multiple roles, as well. It is one of the determinants of the discount rate, and one of the drivers of emissions and hence climate change. We also find that the impacts of climate change grow slower than income, so the effective discount rate is higher than the real discount rate. Moreover, the differential growth rate between rich and poor countries determines the time evolution of the size of the equity weights. As there are a number of crucial but uncertain parameters, it is no surprise that one can obtain almost any estimate of the social cost of carbon. We even show that, for a low pure rate of time preference, the estimate of the social cost of carbon is indeed arbitrary—as one can exclude neither large positive nor large negative impacts in the very long run. However, if we probabilistically constrain the parameters to values that are implied by observed behaviour, we find that the expected social cost of carbon, corrected for uncertainty and inequity, is approximate 60 US dollar per metric tonne of carbon (or roughly $17 per tonne of CO2) under the assumption that catastrophic risk is zero.

**#146** Yohe, G. and Hope, C., "Some Thoughts on the Value Added from a New Round of Climate Change Damage Estimates," *Climatic Change* 117: 451-465, 2013.

This paper offers some thoughts on the value added of new economic estimates of climate change damages. We begin with a warning to beware of analyses that are so narrow that they miss a good deal of the important economic ramifications of the full suite of manifestations of climate change. Our second set of comments focuses attention on one of the most visible products of integrated assessment modeling— estimates of the social cost of carbon which we take as one example of aggregate economic indicators that have been designed to summarize climate risk in policy deliberations. Our point is that these estimates are so sensitive to a wide range of parameters that improved understanding of economic damages across many (if not all) climate sensitive sectors may offer only limited value added. Having cast some doubt on the ability of improved estimates of economic damages to increase the value of economic damage estimates in integrated assessment modeling designed to inform climate policy deliberations, we offer an alternative approach—describing implicitly a research agenda that could (a) effectively inform mitigation decisions while, at the same time, (b) providing economic estimates for aggregate indicators like the social cost of carbon.

# CHAPTER 13

# THE CLIMATE EXPERIENCE
# ON A PERSONAL LEVEL

*The lede: to repeat myself, I always got more out of all IPCC experiences than I ever put in - and I put in quite a bit. I enjoyed the collaborations, but I also enjoyed many experiences that I could not have imagined before I travelled. I had not traveled abroad until after graduate school, but I certainly made up for lost time. Here are a few selected experiences drawn from my experience as a senior member of the IPCC. Some have links to the science and assessment, and others are just stories that show the personal opportunities that my growing involvement in the IPCC created until I pulled back in 2012. For each selected experience from when I became a senior member of the "inner staff", brief description of location and, when appropriate, reference to a paper or a figure is provided.*

## 1993

### Barcelona, Spain

Barcelona hosted the summer Olympics in 1992. It was also the host of the Secretariat of the International Human Dimensions Committee (IHDC) which operated under the direction of our Administrative Officer Christina Poole. Our concern was global change.

I was an executive officer of the IHDC. We held meetings all around Europe. Sometime after the Olympics, we were hosted by the Secretariat in Barcelona. There was, of course, a fancy meal at the end of the first day. Christina Poole introduced me to her friend Christina before the festivities began. Christina Poole then disappeared to tend to the details of the event, and I spent a very pleasant 20 minutes talking to the other Christina. Turns out that she and Christina Poole had won a gold medal for sailing in the recent Olympics representing their host country. I thought that that was very cool.

At some point, Christina Poole decided it was time for dinner to begin, so we should all find our seats. Her friend Christina had a seat at the head table, and I did not. So, it was Europe. I kissed the other Christina on both cheeks; she returned the favor, and off we went. We had had a very nice chat.

Now I am waiting to be served with the appetizer when John Perry, a friend from the US National Science Foundation taps me on the shoulder. He asked,

"Do you know who you were talking to?"

"Yes", I said. "Christina's friend Christina."

"Do you know who she is?"

"Yes. She is an Olympic champion. Gold medal!".

"No no no no… do you know who her father is?"

"I have no idea."

"Her father is the King of Spain. You just kissed a princess and broke 10K rules of protocol."

The evening finished well. Dinner was lovely. As we were departing the hall, the two Christinas came by to say their good-nights. I shoock hands with one Christina for a job well done. I got two more kisses from the other Christina. Apparently being impressed by what she had accomplished and who her friends were was more important than who she was.

## *1997*

### Berlin, Germany:

I am not sure of the precise dates in 1997 anymore, but I do remember the experience. Though it was not exactly an IPCC event, but my invitation to participate in the first joint meeting of the National Academies of what had been East and West Germany was certainly the product of my role in the IPCC. It represented the reunification of an Academy that was created in 1700 as Berlin-Brandenburg.

The conference was held in a new and very fancy Westin Hotel on the west side of the wall. Five stars still, I expect. We were 2 blocks from the Brandenburg Gate, and that was an enormous distraction. On a break, I walked down to the Gate to walk back and forth under its arches. In fact, many times I would sneak away to walk back and forth under the Gate. It was not lost on me that many people had died trying to do what I was doing. It made me cry, and still does. The paper I presented at the invitation of John Schellnhuber is number 48:

**#48.** Yohe, G., "Integrated Assessment of Climate Change – the Next Generation of Questions", *Climate Impact Research: Why, How, and When?"* Akademie Verlag GmbH, Berlin, 2000, Joint International Symposium

## *2000*

### Eisenach, Germany: February 8-11:

**Eisenach** is a town in Thuringia, Germany with 42,000 inhabitants. It is situated near the former Inner German border. Its major attraction is Wartburg castle which has been a UNESCO World Heritage Site since 1999.

Martin Luther came to Eisenach and translated the Bible into German. His room in the monastery was prominently on display during our visit. In 1685, Johann Sebastian Bach was born there.

Ferenc Toth was, and still is, a really really good friend and colleague. He invented "tolerable windows" as a thought framework born of his work in Potsdam. I was more than favorably impressed (see article # 34 below).

Ferenc got really sick in the middle of the night during our meetings in the winter of 2000. He called my room after midnight for help. Disarmed that he would call me, I calmed him down (being really sick in the middle of the night away from home is very scary – trust me).

I spoke with his family through an alarming post-midnight phone call. I called in some favors and found him the next train home; it would stop to pick him up. I had paid for a first class ticket. We collected his papers (the belonging could wait) and I walked him to the train station. It was not a long walk, but it was very dark. It was before dawn. And he was not very mobile.

He had refused going to a hospital in Germany, but that was fine with the family and so with me. I had arranged for his luggage to follow us, so we got everything essential on the train.

He was admitted to a hospital upon his arrival in Austria. There he would for "Eeyeors' – days, weeks, months, maybe even years according to Pooh…... who knows? Ferenc survived, and he is still productive. His and my life perspectives changed that night.

It took me much longer to learn that lesson.

I had seen Martin Luther's room in the nearby church complex earlier in the night during an excursion from a meeting dinner. There was nothing nailed to the door, but it was nonetheless very chilling in its starkness.

**#34** Yohe, G., "Uncertainty, Short Term Hedging and the Tolerable Windows Approach" *Global Environmental Change* 7: 303-315, 1997.


## Antigua and Barbuda: June 4-6:

Antigua and Barbuda is a sovereign island country in the West Indies, lying at the juncture of the Caribbean Sea and the Atlantic Ocean. The country consists of two major islands, Antigua and Barbuda, which are separated by around 40 km (25 mi), and of smaller islands, including Great Bird, Green, Guiana, Long, Maiden, Prickly Pear, York Islands, Redonda). The permanent population number is estimated to be in the region of 97,120 (2019 est.) with 97% residing on Antigua.[4] The capital and largest port and city is St. John's on Antigua, with Codrington being the largest town on Barbuda. Lying near each other, Antigua and Barbuda are in the middle of the Leeward Islands, part of the Lesser Antilles, roughly at 17°N of the equator.

This was a special meeting of the Chapter 18 author team for the IPCC Fourth Assessment Report (Barry Smit, Salem Huq, Ian Burton, et al.).

We all went out for a dinner at the end of a long day. Our hotel mad the recommendation. When we arrived, it was nearly vacant except for a Bride and Groom who were alone in the corner. Ian Burton danced with a new bride on a bet with Barry Smit, but that opened things up. We joined their small wedding celebration. We all left a nice wedding present, they were happy that strangers would celebrate their day when nobody else did. It just seemed right. We all had a wonderful time that closed the restaurant; and they celebrated with strangers from six continents. We danced. We hugged. and

I paid for dinners – ours and theirs. It was expensive, but it felt right. We made them feel special, and they made us special. Since then, I have never paid for a dinner with those guys, and we have had joint dinners on five different continents.

Back to work, I solidified the concept of the "eight determinants of adaptive capacity" with the *Adaptation* chapter 18 author team; it is still an anchor for organizing thoughts about who might adapt positively.

**#57** Yohe, G. and Tol, R., "Indicators for Social and Economic Coping Capacity – Moving Toward a Working Definition of Adaptive Capacity", *Global Environmental Change* 12: 25-40, 2002.

## Lisbon, Portugal: August 8-11:

> Lisbon is Portugal's hilly, coastal capital city. From imposing São Jorge Castle, the view encompasses the old city's pastel-colored buildings, Tagus Estuary and Ponte 25 de Abril suspension bridge. Nearby, the National Azulejo Museum displays 5 centuries of decorative ceramic tiles. Just outside Lisbon is a string of Atlantic beaches, from Cascais to Estoril.

I had dinner with Steve Schneider and others of Steve's IPCC friends at Steve's favorite restaurant on Lisbon's Atlantic coast. We had giant prawns, and that was it. Just outside the restaurant, fishermen would bring their day's catch onto the beach in afternoon. You could pick you dinner fish or prawn and take it to a local restaurant for preparation. Steve selected our prawns and made our reservation. They were big. Two were sufficient and we all had three.

Getting back to work, a big fight erupted at the authors' meeting between ecologists and economists. What can we say about attribution to climate change for observed impacts on natural species? That was the question. People were screaming. Others were crying.

Conferring in the back of the with Camille Parmesan, whom I did not know, we started to talk. She was an ecologist and I was an economist. We represented the argument's opposing sides. But we gathered our thoughts about how to approach common ground. As a result of our conversation, Lisbon became the birthplace of the Parmesan and Yohe (2003) paper in *Nature*. I cannot speak for Camille, but this is my most cited paper, at 15K+ in early 2023.

This experience, and the subsequent collaboration, is perfect evidence of "getting more out of IPCC than you put in, even if you are not paid". We got our first referee report from Steve Schneider. He was also hanging out at the back of the room. I noticed, and knew that this was troubling for him. Camille was reluctant because she did not know Steve and was worried about whether he would be an honest broker. I prevailed on be basis of my friendship. We approached Steve and described our idea. He listened for many minutes, interrupting as always with questions. Then he said "This sounds like a paper for *Nature*." Though it took a lot of work, he was right.

I also played golf with Bill Easterling on a course near the ocean that regularly hosts the Portugal Open – then, a regular stop on the European tour. I shot 74, so I would not have made the cut.

**#60** Parmesan, C. and Yohe, G., "A Globally Coherent Fingerprint of Climate Change Impacts across Natural Systems", *Nature* 421, 37-42, January 2, 2003.

## *2003*

### Colombo, Sri Lanka: March 5-7:

**Colombo** is the executive and judicial capital and largest city of Sri Lanka by population. According to the Brookings Institution, Colombo metropolitan area has a population of 5.6 million, and 752,993 in the Municipality. It is the financial center of the island and a tourist destination. It is located on the west coast of the island and adjacent to the Greater Colombo area which includes Sri Jayawardenepura Kotte, the legislative capital of Sri Lanka, and Dehiwala-Mount Lavinia. Colombo is often referred to as the capital since Sri Jayawardenepura Kotte is itself within the urban/suburban area of Colombo. It is also the administrative capital of the Western Province and the district capital of Colombo District. Colombo is a busy and vibrant city with a mixture of modern life, colonial buildings and monuments.

I stayed in another Intercontinental Hotel – this time in Colombo. Sri Lanka was a long and arduous trip. I had to go through three layers of security to get to a plane at the end of a remote terminal in Frankfurt. When I was finally there the next day, Sri Lanka was in the middle of a civil war. A town bus had recently been bombed next to the capital building in broad daylight – one block from the hotel in which I was staying; and the remains of the bus were still there to remind someone of something.

The hotel offered tours of the capital, but that suggestion received a very quick "no, thank you."

I did take a day-long country excursion outside the city limit - seeing elephants and cobras and Buddha temples and bombed out busew with Richard Moss and two other colleagues. We did not recognize the danger of the trip until we saw the carcasses of more buses on the way back to the hotel; they were still burning.

Obviously, we were there during a civil war. White guy from New England... what could possibly go wrong.

During a coastline walk near the hotel along the beach to the Indian Ocean, I witnessed the arrest of somebody by heavily armed soldiers. The episode began with my being stopped by a soldier. "Let me see your papers". Happily, I had mine with me. When a single walker approached from the opposite direction, he added "maybe you should stand behind that shack." The shack was constructed with ¼ inch plywood. "How about that stone wall?", I replied. He nodded and I retreated behind the stones when he agreed. I left my passport and other papers with the soldier, of course, so.....

Six or seven armed soldiers suddenly appeared from nowhere. They pointed their machine guns at the approaching walker who acquiesced. When they took their hostage away, my soldier friend (as he was now) signaled for me to come out. "Here is your passport and your papers. Why don't you go back to the hotel? NOW?"

My flight home left at 3AM two days later, but that was my last outside escursion. At that time or of the moring, highway markings and laws were suggestions, and my hotel driver played 'chicken' with all comers all of the way to the airport.

I was there for seven days. We were served curry all the time - even eggs for breakfast tasted like curry. I liked curry when I arrived, but not so much when I left. The Beachboys' "This is the worst trip I've ever been on" still plays in my head whenever I think of Colombo.

I do not remember what we accomplished.

## *2004*

### Maynooth, Ireland: May 18-20:

> **Maynooth** is a university town in north County Kildare, Ireland. It is home to Maynooth University (part of the National University of Ireland and also known as the National University of Ireland, Maynooth) and St Patrick's College, a Pontifical University and Ireland's sole Roman Catholic seminary. Maynooth is also the seat of the Irish Catholic Bishops' Conference and holds the headquarters of Ireland's largest development charity, Trócaire. Maynooth is located 24 kilometres (15 miles) west of central Dublin.

I do remember what we accomplished in Maynooth. A small group of IPCC authors selected from all three working groups were gathered to draft uncertainty guidance for the next assessment. We made progress, but the product did not stand the test of time.

The day after the sessions and before my very early departure, I took a daylong excursion with Camille Parmesan to the nearby pyramids and Dublin. We had a great and productive great until well into the night in the city. We kept running into Roger Jones all day, but we left him in Dublin.

Camille and I were still working on the *Nature* paper. We made some progress. I also spent a lot of time in my hotel room working on the modeling and text for the hedging paper with Natasha Andronova and Michael Schlesinger that ended up in *Science*.

Not a bad week for an economist.

**#61** Yohe, G., Andronova, N. and Schlesinger, M., "To Hedge or Not Against an Uncertain Climate Future", *Science*, 306: 415-417, October 15, 2004.

**#62** Manning, M.R., Petit, M., Easterling, D., Murphy, J., Patwardhan, A., Rogner, H-H, Swart, R. and Yohe, G. (eds), IPCC Workshop on Describing Scientific Uncertainties in Climate Change to Support Analysis of Risk and of Options: Workshop report. Intergovernmental Panel on Climate Change (IPCC), Geneva, 2004.

## *2005*

### Capetown, South Africa: May:

> **Cape Town** (Afrikaans: *Kaapstad*; [ˈkɑːpstat], Xhosa: *iKapa*) is one of South Africa's three capital cities, serving as the seat of the Parliament of South Africa. It is the legislative capital of the country, the oldest city in the country, and the second largest (after Johannesburg). Colloquially named the *Mother City*, it is the largest city of the Western Cape province, and is managed by the City of Cape Town metropolitan municipality. The other two capitals are Pretoria, the executive capital, located in Gauteng, where the Presidency is based, and Bloemfontein, the judicial capital in the Free State, where the Supreme Court of Appeal is located.

Bob Kates, a dear friend and incredible colleague, got very sick from dehydration in Cape Town. He recently passed away, in May of 2018, but his contribution in Capetown is stil remembered.

Before he was taken to the hospital, Bob served as our review editor on the zero-order draft of #86 for the first morning of the meetings. He broke the rules of non-engagement by agreement with me; I was Co-ordinating Lead

Author with connections so, as we began our authors' first discussions as a group, I asked him for his thoughts. He had told me what he would say, and so I asked him to speak when our meeting began.

He said something like "you are all very smart, so stop trying to show everybody that you are very smart. Start over and write something that might actually be useful". I started the resulting what would become a thorough revision process by tearing up my 40% submissions to the zero-order draft and throwing them on the floor. Everyone followed. We started over, and our chapter was much better for it.

I gave the chapter team an afternoon off on the third day off so that I could take a countryside excursion with Chris Hope. We saw the country, the coastline along with seals and penguins. We also saw some small communities, and we had dinner in a small grill in one of them – very good, but no recollections of which establishment and where.

We also drove through shanty-towns. I learned that you can book a week in a shanty. I could not decide whether or not that would be a good idea, but I have thought about it since.

*2006*

### Merida, Mexico: January 16-19:

> **Mérida** is the capital of the Mexican state of Yucatán, and the largest city in southeastern Mexico. The city is also the seat of the eponymous Municipality. It is located in the northwest corner of the Yucatán Peninsula, about 35 km (22 mi) inland from the coast of the Gulf of Mexico. The city's rich cultural heritage is a product of the syncretism of the Maya and Spanish cultures during the colonial era. It was the first city to be ever named American Capital of Culture and is the only city that has received the title twice. The Cathedral of Mérida, Yucatán was built in the late 16th century with stones from nearby Mayan ruins and is known to be the oldest cathedral in the mainland Americas. In addition, the city has the third largest old town district on the continent. In 2007, the city was visited by former U.S. President George W. Bush to meet with the ex-president of Mexico Felipe Calderón for the historic creation of the Mérida Initiative.

Linda and I traveled here with Bob and Joan Wilson for the second authors' meeting of AR4 Working Group II. Bob and I were old grad school friends. He knew nothing about climate change. Linda and Joan were old friends, and that helped. We four happily took multiple day trips together to pyramids and other sights before the meetings started. We offered Texas to our driver. He declined.

We also all attended an IPCC official dinner with young Mexican children singing along a candle-lit path from the meeting hotel to the US consulate, for a reception – all hosted by Mexico. We all ate at the table with IPCC Chair Pachauri and a very few others. Joan chatted Pachauri up, and he was very happy. This was a highlight of a wonderful trip.

It was a Working Group II authors' meeting. Our final product would be:

**#86** Yohe, G., R.D. Lasco, Q.K. Ahmad, N. Arnell, S.J. Cohen, C. Hope, A.C. Janetos and R.T. Perez, "Perspectives on climate change and sustainability", in *Climate Change 2007: Impacts, Adaptation and Vulnerability. Contribution of Working Group II to the Fourth Assessment Report of the Intergovernmental Panel on Climate Change* (M.L. Parry, O.F. Canziani, J.P. Palutikof, C.E. Hanson and P.J. van der Linden, eds.), Cambridge: Cambridge University Press, 2007.

## Geneva, Switzerland: August 1-3:

Geneva is Geneva, but there is always a story there; read on. Just for reference, it is a city in Switzerland that lies at the southern tip of expansive Lac Léman (Lake Geneva). Surrounded by the Alps and Jura mountains, the city has views of dramatic Mont Blanc. Serving as headquarters of Europe's United Nations and the Red Cross, it's a global hub for diplomacy and banking.

I chose a new hotel for this Geneva visit (having been there many times); that was a habit of mine, since I was had been there so often and chosing something closer to the meeting venue was becoming attractive.

This time, I chose one that was walking distance to the UN office building where the meetings were to be held. It turned out that it was across a very small street from the back of the Libyan Embassy. It was very popular for guests from Muslim countries, and that was fine with me.

It was welcoming to me, and its restaurants served great food. Geopolitics were tense in 2006, but the local environment and the food was not.

On the third day of the writing meetings, late in the afternoon, I was walking alone back to my hotel on the other side of the street. I was planning what I would write that night for tomorrow's final session after a quiet dinner. That is what I did on these trips.

During my walk, I could look through the windows that exposed the lobby of my hotel. I could see diverse families enjoying early meals together, and that made me smile. I was, though, seriously distracted

As I was looking back to my left to the view of the lobby, the a small and hitherto nondescript back door to the Libyan Embassy suddenly opened in front of me. I ran right into Condoleezza Rice. She was then the Secretary of State of the United States. Quite literally, I had bumped into her. Actually, I almost ran her over. We glanced at each other with smiles, but then her people arrived.

She was coming out of the nondescript back door of Libyan Embassy where her car was waiting. It was nearly dark. It turns out that she was "not supposed" to be there. The US did not have diplomatic relations with Libya, but the Libyans did have back channels with the Palestinians.

I was quickly approached by somebody with a communication device in his ear while she rushed into her car. We were surrounded by others with many threatening bulges – most had ear-devises and some had some sort of serious weapons at the ready.

My initial serious person asked for my papers. Happily I had them with me.

He asked what I had seen. I said "Nothing, sir." He smiled.

He gave me back my papers as he asked what I was doing in Geneva. He went to his place in her car and the posse dispersed. While he was walking to the car, I asked that he give my best wishes to the Secretary on her mission on whatever she was working on.

"The world needs her to succeed", I said. He smiled. Across the street, the Libyans smiled, as well. End of story.

At the authors' meeting in the new UN headquarters building, the next day we were working on the final "first draft" of the Synthesis Report for the entire Fourth Assessment.

Things were getting very tense from the second day of the meeting. Perspectives from developed and developing nations had emerged, and so I thought it would be good to break the tension.

I created a Summary for Policymakers (17 words and not 22 pages) in the form of a "haiku" portrayed against the IPCC image for the cover page of the entire report. It said:

*Climate is changing*
*Humans are to blame*
*The poor will suffer most*
*The rich don't care.*

I showed it to Pachauri, then the Chair of the IPCC. He suggested an alternative last line through a scribbled note on a napkin of all things:

**The rich don't give a damn.**

Not quite a haiku after his revision, but the plenary got to pick its version knowing that neither would be on the cover.

Patchy put his version of his slides for all to see. Framed visually in what could be presented on the become the cover of the Synthesis Report that we were all working on.

He offered to call it a day and publish the modestly revised version that emerged from the subsequent deliberations. I applauded, and we made eye contact with Patchy. We smiled.

Another eight months of writing produced our report, but at least the tone of the language recognized that Geneva event.

**#87** Bernstein, L., Bosch, P., Canziani, O., Chen, Z., Christ, R., Davidson, O., Hare, W., Huq, S., Karoly, D., Kattsov, V., Kundzewicz, Z., Liu, J., Lohmann, U., Manning, M., Matsuno, T., Menne, B., Metz, B., Mirza, M., Nicholls, N., Nurse, L., Pachauri, R., Palutikof, J., Parry, M., Qin, D., Ravindranath, N., Reisinger, A., Ren, J., Riahi, K., Rosenzweig, C., Rusticucci, M., Schneider, S., Sokona, Y., Solomon, S., Stott, P., Stouffer, R., Sugiyama, T., Swart, R., Tirpak, D., Vogel, C., and Yohe, G., *Climate Change 2007: Synthesis Report (for the Fourth Assessment Report of the Intergovernmental Panel on Climate Change)*, Cambridge: Cambridge University Press, 2007.

## *2007*

## Windsor, United Kingdom: Authors' meeting for the Technical Summary (TS) of the Working Group contribution to the AR4

Windsor is a historic market town and unparished area[1] in the Royal Borough of Windsor and Maidenhead in Berkshire, England. It is the site of Windsor Castle, one of the official residences of the British monarch. The town is situated 21.8 miles (35.1 km) west of Charing Cross, central London, 5.8 miles (9.3 km) southeast of Maidenhead, and 15.8 miles (25.4 km) east of the county town of Reading. It is immediately south of the River Thames, which forms its boundary with its smaller, ancient twin town of Eton. The village of Old Windsor, just over 2 miles (3 km) to the south, predates what is now called Windsor by around 300 years; in the past Windsor was formally referred to as New Windsor to distinguish the two.

Our meetings were held on a lovely campus within walking distance of Windsor Castle. Good for a break. I joined some friends to take a walk to the Castle.

This was a final authors' meeting for the Technical Summary (TS) of the contribution of Working Group II to the AR4. The big news for me was that a first year student of mine, Amanda Palmer, made the first cut. And the last cut. Amanda Palmer, an 18 year old soon to become a Sociology major, thereby became a contributing author to chapter 18 and so, by virtue of the AR4, a 2007 Peace Price Nobel Laureate.

Let me explain what happened. Her insight, from work she did for a First Year Initiative course at Wesleyan (mine) when I was in Capetown, was that climate impacts seem to pile up along the southeastern coast of Africa, among other places. As a result, their impacts should be expected to compound – Figure 13-1 below was her primary visual.

That is, she argued that sum of the compounded impacts of different manifestations of climate change experienced at the same time in the same place was likely to be larger than the sum or their separate parts taken one at a time. If her hypothesis were true, then adaptation to climate impacts would be much more difficult and would need to be coordinated across multiple impact vectors.

I walked her map and her hypothesis around the room near Windsor at the TS authors' meeting. It was a new insight to most of my author friends, and they agreed that her hypothesis was robust.

All was not well when the AR4 released this finding. Amanda was called out by the dismissive fellows at the Heartland Institute for having no credentials. James Taylor wrote something like: "[h]ow can you believe anything from the IPCC if she (a first student at a liberal univesity is a contributing liberal New England school. God only knows what they teach there, and then they send if around the world. Everything in the report should be viewed with extreme skepticism."

We at Wesleyan teach things like "have a look at this embargoed text of the Africa chapter draft and find something new". She did, and she was right.

This entire story is the product of a singular and significant episode from a fall seminar course that I had agreed to teach to first year students in their first semester in Middletown. My students were self-selected incoming students who were interested in the environment and therefore attracted to climate change. I had told them that I had to attend the Capetown IPCC authors' meeting during their first week of class. I had told them what IPCC was, and reminded them how far away Capetown was – go to Frankfurt and turn right for 10 hours.

During the week that I would be gone, I asked them all to find something of interest about climate change, working with a librarian who had agreed to cover for me. Learning about how to work in a world class college library during their first week in college could not hurt, I had rationalized. The assignment came with a catch: they were to give presentations of what they found in the literature that they had found when I returned.

I expected that I would return to "high school book reports" of articles from places like *Scientific American* and maybe the *Wall Street Journal.* I had shared some drafts of Working Group II chapters as a tease for what we were going to cover later in the fall just in case somebody would want to present something from that reading without going to the library. It was appropriate on my part because the IPCC drafts had been sent were for public review, and my 15 students were part of the public. Surely somebody, I thought, would summarize their reading of an IPCC chapter in 10 minutes.

Welcome to the Wesleyan classroom. NEVER try to teach there with a cold, and NEVER walk into class overprepare. You NEVER know what to expect.

### Estes Park: July 31 – August 3:

> **Estes Park** is a statutory town in Larimer County, Colorado, United States. The town population was 5,904 at the 2020 United States Census. Estes Park is a part of the Fort Collins, CO Metropolitan Statistical Area and the Front Range Urban Corridor. A popular summer resort and the location of the headquarters for Rocky Mountain National Park, Estes Park lies along the Big Thompson River. Landmarks include *The Stanley Hotel* and The Baldpate Inn. The town overlooks Lake Estes and Olympus Dam.

**This one is important.** It was a meeting of the core writing team for the Summary for Policymakers of the AR4 Synthesis Report. We were hosted by the Stanley Hotel of Jack Nicholson (*The Shining*) fame. There is even a Stanley Steamer in the lobby.

Susan Solomon and Bert Metz chaired in the absence of Martin Parry. Stephen Schneider and I worked together with William Hare long into the early morning on many nights, so that I (we) can confirm that this was the birthplace of the IPCC language about "iterative risk management" - "Responding to climate change involves an 'iterative risk management' process that includes both adaptation and mitigation and takes into account climate change damages, co-benefits, sustainability, equity and attitudes to risk." This is the language that has been explored more thoroughly in chapter 9.

I remember working with Steve early into the morning of one night in between days 2 and 3. He suddenly asked to see what we had produced over the previous 3 hours. My computer had just frozen, and so I could not show him. At least not immediately. I was suddenly afraid that 3 hours of our joint work had disappeared into the ether.

I remembered, though, that I had been taught about the value always saving early and often. All was not lost. Our work was on my flash drive. Phew!

### Brussels, Belgium: April 2-5:

> Brussels (Belgium) is considered the *de facto* capital of the **European** Union, having a long history of hosting a number of principal EU institutions within its European Quarter. The EU has no official capital, and no plan to declare one, but **Brussels** hosts the official seats of the **European** Commission, Council of the **European** Union, and European Council, as well as a seat (officially the second seat but de facto the most important one) of the European Parliament.

This one is important, as well. It was the plenary meeting of UNFCCC member nations to approve the contribution of WGII to the AR4. I was presenting and defending language before the plenary during a session that lasted until well past midnight. It was then when our assessment of the *Stern Review* was presented for consensus.

We had asserted over and over againe in the face of opposition from the UK delegation that the *Stern Review* (which had been commissioned from the highest levels of the British government) had not been peer reviewed. Finally, David Warlow from the UK delegation that was there in force despite the late hour, asked "How do you know that?" "How do you know that the *Review* was not peer reviewed".

I was reminded, immediately, of the lesson for every law student in his or her first year - Never ask a question for which you do not know the answers.

My answer, around 2AM in plenary was clear: "I know because Sir Nicholas told me so at a Yale event several weeks ago" That ended that kerfuffle, but...

Debate over a "high confidence" assessment by the authors about impacts on ecosystems also erupted with Saudia Arabia, China, Kuwait; they wanted, at most, "medium confidence." IPCC authors were solid with their assessment of at least "high confidence."

IPCC authors are, however, relegated to the "children's table" in these proceedings. By rules from grandmother or the IPCC Secretariat, they cannot speak unless they are spoken to.

This was a problem, except that some of us let it be known to our Country delegations that IPCC rules said that authors' assessments could always be included as footnotes if they disagreed with what plenary has decided. That rule had always been on the books since day one of the IPCC Charter, but it had never been tested. We (authors) and they (country delegates) all understood that exercising that clause would damage the entire IPCC process.

Many authors walked out and threatened a footnote when certain countries continued to object to the content of a finding in the text that human actions were (with high confidence) demonstrability damaging natural systems and specific species. One author of an ancillary chapter exclaimed on his way out the door that resistance to this finding amounted to "intellectual vandalism" by people who did not know what they were talking about.

This all also happened in the middle of the night. Stephen Schneider understood the significance of this dispute and the protest. He had also been involved in drafting the offending finding. Steve and I finally suggested to the plenary that no confidence statement be attached to the finding after Cynthia Rosenzweig (the Convening Lead Author) walked out. We came up with a plan: take the confidence statement out. The authors agreed, but we could not get it to the floor for the nations' consideration because we were authors. So....

I spoke offline with Trig Talley from the US delegation. Would the Bush White House to agree with this compromise? If it did, would the US put the proposal forward and engage with China?

Trigg arranged a **very** early morning (U.S time) conference call with the President (of the United States), who agreed. Trigg then went to the Chinese delegation to get their approval, and China brought everyone else along.

There would be no footnote, but everyone would know that something had happened because no confidence statement would be a glaring omission. And the conclusion that caused so much trouble? Have a look:

> Observational evidence from all continents and most oceans shows that many natural systems are being affected by regional climate changes, particularly temperature increases.

The next finding was:

> A global assessment of data since 1970 has shown it *is likely* that anthropogenic warming has had a discernible influence on many physical and biological systems

The contrast of assessment language with and without a confidence assessment (my emphasis) was visible by direct comparison.[31] The evidence for both is displayed in Figure 13-2.

**Valencia, Spain: November 12-17:**

> **Valencia** is the capital of the autonomous community of Valencia and the third-most populated municipality in Spain, with 791,413 inhabitants. It is also the capital of the province of the same name. The wider urban area also comprising the neighboring municipalities has a population of around 1.6 million, constituting one of the major urban areas on the European side of the Mediterranean Sea. It is located on the banks of the Turia, on the east coast of the Iberian Peninsula, at the Gulf of Valencia, north of the Albufera lagoon.

Valencia in November is where and when plenary consensus approval of the Summary for Policymakers for the AR4 was accomplished, including the 30 words that put iterative risk management before the world's decision-makers (chapter 9).

After the last session, which was an enormous success, Pachauri invited me to come along with him and a few others for a dinner and nighttime tour hosted by the President of Valencia. We would be celebrating the surprising

---

[31] Source: pages 8 and 9 in https://www.ipcc.ch/site/assets/uploads/2018/02/ar4-wg2-spm-1.pdf.

result of complete success in that the "iterative risk management" language had been accepted by consensus. We knew that it would, at the very least, be the foundation for subsequent UNFCCC negotiations at the Conferences of the Parties.

After a spectacular dinner, we drove around town with police escort. The President showed us the sites of his beautiful city.

After midnight, we stopped at the Opera House, a magnificent new structure that had exploded onto the world stage with its outstanding acoustics. The President had a key to the building and every room inside, so in we went for a tour. Every room, including the theater, was dark until the lights would magically come on when he announced himself.

We ended up in a narrow hallway with office doors and little else. The President walked down the hall for a bit and then knocked on one of them with light leaking out from its bottom. That is how I met Zubin Mehta. He was there working on an upcoming performance. He was very gracious, despite the interruption. I was overwhelmed.

Thanks to that experience, I hand-imported 3 bottles of Valencian wine back to Connecticut. They were a gift from the President. I had found in my hotel room when I finally returned from the tour, with not much time to prepare for my flight home – an arrangement made with security so that I would be not lugging the bottles of wine all through the night (and through security at the airport).

I have been blessed to have had some spectacular experiences around the world. More on that later, but this whole week was one of them. We had changed the world, and my wine got home to Connecticut.

*2009*

## Venice, Italy: (Scoping Meeting for AR5): July 13-17:

**Venice** is Venice - a city in northeastern Italy and the capital of the Veneto region. It is built on a group of 118 small islands that are separated by canals and linked by over 400 bridges. The islands are in the shallow Venetian Lagoon, an enclosed bay lying between the mouths of the Po and the Piave rivers (more exactly between the Brenta and the Sile). In 2020, around 258,685 people resided in greater Venice or the *Comune di Venezia*, of whom around 55,000 live in the historical island city of Venice (*centro storico*) and the rest on the mainland (*terraferma*).

The fun part first. I got to hand-import a glass sculpture to home because I played hooky to buy it. We were meeting on an island, so an excursion to the shopping involved a boat. That day, I met several colleagues from the meeting who were doing the same thing. The sculpture is still in our living room, but that is not the best part of the story of that excursion.

There was a failed "terrorist attack" at San Marco Square when I was there making my purchase. Three men had climbed the clock tower with machine guns and shouted down at the square something that I did not understand. They were menacing. They were heavily armed. Gut nobody paid any attention. After about 10 minutes of being ignored, they packed up their weapons, climbed down the tower from its very top, and escaped without being apprehended.

At the meetings, Rob Mendelsohn insisted that maximizing benefit-cost differences (B-C) was the only way to do the economics of climate change. Many times, he made that point to deaf ears The last time was in a working group meeting in front of Pachauri, who had shepherded the iterative risk management approach through a plenary not more than 12 months before. I was convening the meeting. I made eye contact with Patchy just before he left. The message was clear, and I agreed. Forget about it!

Rob persisted. He asked "Why not commission an IPCC Special Report on B-C?"

Charlie Kolstad, our chair, correctly resisted Rob's position, but it had some serious support in the room. I proposed tabling the idea until the next day so that I (maximum IPCC experience) could prepare a presentation about exactly how much work would be involved in producing such a report under the same IPCC rules as any other assessments - multiple drafts, public and government review, redrafting, and a plenary to achieve consensus word by word.

The next day, the group agreed that writing a Special Report on B-C would be a waste of time.

During the initial plenary, the Chair of Working Group I (Thomas Stocker) said that he would not allow his assessment to include a single paper with a unique or contradictory finding. Since the UNFCCC, and much of the world, had accepted risk management as the approach for response, this struck me as heresy. Specifically Working Group I was dismissing the insights and influence of climate change on humanity.

Pachauri recognized me for an intervention immediately after Chair Stocker made that statement. Then and then, I said something like "Your words made the hairs on the back of my neck stand on end. Your approach would immorally eliminate 10 years of IPCC progress". I was annoyed, to say the least

In the silence that followed my ranting, Pachairi recognized a two-handed intervention from Steve Schneider. He began "What Gary meant to convey was, ……..". Steve had my back. He rephrased my comments so that they were less toxic. He thereby reversed our productive history. I smiled, because it confirmed that I was not also right, but also that I had finally graduated from his school on activism.

On a happier and funnier note, Chris Field, the Working Group II Chair, had invented an abbreviation code to indicate the potential for chapter cross cutting themes to be included multiple chapters so that collaborative confidence (or lack thereof) would be clear.

I had missed the explanation of the code because of my shopping excursion, so I was confused when Chris presented drafts across portions of the scoping outline with to which he had attached a collection of capital letters that he thought had some meaning. They did, but not to me. When he got to the economics part, I noticed that he had attached "CSTDRMPT" to the Chapter. I asked, verbally in plenary, "What does 'costed armpit' mean?"

Kris Ebi, sitting behind Chris on the stage since she was Chair of the Technical Support Unit, almost fell off the stage in laughter at the question. Lots of people laughed. Chris gave me a perplexed look, the intensity of which was diminished by the levity in the room. Nobody really understood the point of the letters code that he had created, and I had broken the ice much to their relief.

The code persisted in the writing process, but only as "don't forget about" suggestions – much to the benefit of Working Group II's ultimate contribution to the AR5.

## *2010*

## Jasper Ridge, California ((IPPC): July 14/16:

The **Jasper Ridge Biological Preserve** is a 483 hectares (1,190 acres)[1] nature preserve and biological field station formally established as a reserve in 1973. The biological preserve is owned by Stanford University, and is located south of Sand Hill Road and west of Interstate 280 in Portola Valley,

San Mateo County, California. It is used by students, researchers, and docents to conduct biology research, and teach the community about the importance of that research.

The task of a few selected authors from each IPCC working group and other invited experts was to prepare an updated uncertainty guidance document for all IPCC assessment authors. This was to be guidance for the upcoming AR5.

We finally got it right. These few words hav been applied everywhere since then, including the AR6, the fifth US National Climate Assessment, and so on.

The meetings at Estes Park were the last time I would see Steve. He was very ill, but he was still traveling the world to give talks and engage skeptics. Terry Root, his wife, was worried. So were we all.

He had promised that he had one more trip and that that would be his last (for a while, which we all understood).

We all enjoyed a lovely meal and some community time at their house. We were all calm and comfortable. The dogs were welcoming. None of us knew that he would die, soon. That is why it hit so hard within a week or two.

In his initial presentation, Chris Field (our host and then still Chair of Working Group II) had warned us about four local hazards at Jasper Ridge: fire (there was evidence), cougars (there were night pictures), snakes (there were specimen pictures), and poison ivy (that was obvious even on the way in).

As a result, none of us strayed very far from the meeting cottage during coffee breaks. Chris did take us out to the site of his field experiments, and that was fun.... Except that Kevin Trenbreth fake-pushed me into some weeds as we viewed the site out in the "wild". Not funny, Kevin.

## *2011*

### San Francisco, California ((IPCC): December 12-15:

**San Francisco** is San Francisco just like Venice is Venice. It is a commercial and cultural center in Northern California. The city proper is the 17th most populous in the United States, and the fourth most populous in California, with 815,201 residents as of 2021. It covers a land area of 46.9 square miles (121 square kilometers) at the end of the San Francisco Peninsula making it the second most densely populated large city in the U.S. Among the 331 U.S. cities proper with more than 100,000 residents, San Francisco has ranked first by per capita income.

This was another authors' meeting. It is memorable for two things, as well as the boat ride under the Golden Gate Bridge at night.

First, I learned something very surprising during a conversation over drinks with California Governor Jerry Brown. He had attended the IPCC reception because he was interested in our work, and he related the following Ronald Reagan - IPCC story:

He told us that the Intergovernmental Panel on Climate Change was "intergovernmental" because President Reagan did not want the scientific arm of the UNFCCC to pick up all of the UN baggage that would go along with being a new UN entity. The very conservative President of the United States had correctly discerned that direct connection of the IPCC with the United Nations would damage its credibility at just the time and place in the process its credibility could be most important. In other words, Ronald Reagan was worried about the perceived credibility of science. I taught this story many times – never mistrust the personal commitment of political leader to the public good. Before I retired in 2019, I stopped teaching that lesson.

This meeting was also the birthplace of detection/attribution confidence matrices. I designed them not only for their visual appeal, but also because every chapter team could create one. Collectively, they would thereby create a collection of comparable images across sectors and regions. It took some work, but the approach was adopted in plenary, and most chapter author teams agreed to give it a shot – see Buenos Aires below.

## 2012

### Tsukuba, Japan: January 11-14:

**Tsukuba** is a city located in Ibaraki Prefecture, Japan. As of 1 July 2020, the city had an estimated population of 244,528 in 108,669 households and a population density of 862 persons per km². The percentage of the population aged over 65 was 20.3%. It is known as the location of the **Tsukuba Science City** *(Tsukuba Kenkyū Gakuen Toshi)*, a planned science park developed in the 1960s.

On the way to Japan, Chris Field and I presented climate change risks to Bill Gates in Seattle on January 7th. David Keith had arranged the meeting;

he was working with the person with whom he had collaborated to write the original Excel code for Microsoft (Charles Simonyi).

Mr. Gates had asked in December for a meeting in January, and he gave us the entire month of January as a window for scheduling our visit. He asked for papers to read in preparation; Chris and I both sent 10 academic papers for his perusal. We expected that, when we got the meeting scheduled for 1.5 hours, we would be giving presentations of those papers – synopses of content and why we sent them to him. At the beginning of the meeting, Mr. Gates made it clear that he had done his reading. He had some questions.

When our discussion neared the 1.5-hour mark, Mr. Gates asked if we could stay longer. We said yes; our plane to Norita was much delayed usual in these connecting days to Asia. It had happened several times in both of our experiences. In fact, Mr. Gates extended the actual meeting by 2.5 hours and we both made our flights to Norita in Japan.

For what ended up being a meeting extended by 4 hours, I sat next to Mr Gates. I saw the questions that he had scribbled on his printed text of the papers that Chris and I had sent for his homework reading. The session felt like a final oral exam, except that we had defined the reading list even though the questions were left to him. At least I could anticipate his questions to me by glancing at his notes ahead of time.

Mr. Gates concluded that climate change was the first issue that he had ever confronted for which technology was not the solution. The room went silent (the two of us and ten of his long-time colleagues).

The suffering would continue, he said, to be more explicit. Colleagues from his team would later admit to us as we made our way to the airport that this was the first time in their experience with him where he admitted that technology could not fix a problem.

I don't remember what we did at the authors' meeting in Japan, but I do remember that we had to take a late-night bus to Tsukuba. It took more than an hour and my plane was late, so it was early morning when I checked into the hotel.

I had been placed in a close and very nice hotel even though I had arrived on a later than expected flight (by many hours). My friend to this day, Rich Richels, was waiting up late at the bar. I shared a drink with him and several other jet-lagged colleagues. Three days later, Rich would arrang for me to be upgraded to a first class ticket on my flight home – that was so cool.

## Lima, Peru (Intergovernmental Panel on Climate Change): June 23-26th.

It was an experience to reach my hotel where we all were billeted.

BUT, we all took motorcades from the hotel to the meetings in government buildings. I found a pen from Hillary Clinton in a drawer of a table in the middle of a big conference room in the Foreign Ministry building that had been reserved for the higher us participants. She had been in Peru the week before as Secretary of State for the United States, and had apparently left the pen as she left.

I presented to the conference on the economics of iterative risk management to an expert meeting on costing and ethics. There, there were deaf ears. There, though, I met Geoff Heal. He was and still is a high-powered US economist. He was not convinced walking into the meetings, but he was when he left – not just from me, but because he finished conversations with experts on the ground. Good for him.

On a personal note, the only quick take way from a Five Star Hotel in downtown Lima is that traveling to the airport in the middle of the night through residential communities – coming and going late at night and very early in the morning – is the only way to come and go. Sorry for the people who live there/

*2012*

## Buenos Aires, Argentina (IPCC): October 23-26:

The **Autonomous City of Buenos Aires** is the capital and primate city of Argentina. It is located on the western shore of the Río de la Plata, on South America's southeastern coast. "Buenos Aires" can be translated as "fair winds" or "good airs", but the former was the meaning intended by the founders in the 16th century, by the use of the original name "Real de Nuestra Señora Santa María del Buen Ayre", named after the Madonna of Bonaria in Sardinia, Italy. Buenos Aires is classified as an alpha global city, according to the Globalization and World Cities Research Network (GaWC) 2020 ranking.

I had previously offered a template of a figure template designed to reflect authors' confidence in detection and attribution. The thought was that comparable representations of each chapters' assessments of their understandings of connections between the detection and attribution (in their region vis a vis a global aggregate) would help their readers understand any regional hesitancy to accept potential attribution of a detected global climate change to local climate drivers.

I hosted a private dinner for *all* of the chapter authors who had accepted the challenge of adopting the figure. Each would ultimately confirm that they would continue to employ it in the D&A sections of their chapters. The hope of those from above was that their doing so would positively inform the message of overarching synthetic chapters.

Meanwhile, the dinner was success. The idea had always been (since IPCC does not set directives from above) for as many chapters as possible to accept the visual as useful and so use it in their required contributions on D&A. Each participating chapter would thereby reveal their summary in a widely applicable visual.

Preparing such a figure in each chapter was not free. Members of author teams had different ideas, and so the creation of each figure involved an enormous amount of discussion; but I thought that those conversations should be

To commend those who undertook this challenge, the dinner recognized a collaboration of authors that were willing to make the effort to incorporate the matrix into their mandatory sections on detection and attribution. Authors from more than 50% of the chapters attended, but the percentage that used the matrix was much higher than that - if only they had recognized the invitation to a free meal. The template would ultimately permeate their chapters as support a consistent synthetic representation in both the Technical Summary of WGII AR5 and its Synthesis Report.

More than 40 authors from more than 20 chapters and five continents attended the dinner. All were grateful for the meal and the collegiality. We all moved from table to table all night to get to know each other - who was there and who was doing what with respect to detection and attribution.

The meal lasted more than 4 hours. Given common interest, the event itself lasted about 5 hours. I paid for it all (even the extra time), and the restaurant was pleased to be part of IPCC history. I am sure that memories of that meal continued to provide motivation the next time IPCC leadership asked all author teams to do something to promote comparability across chapters.

Of course, it was expensive – a fancy restaurant 10 blocks from the hotel. We all walked there as a group for protection, and that the table seating was mixed since most did not know each other. It was not segreatated by continent; was the best I could do to foster friendly engagement on the way to and during dinner.

Happily, and before I went to Argentina, Wesleyan had agreed to let me use my private research money to cover the expense; they had the authority to allow me to use my name to cover the expenses. They were happy to have their name on the occasion and did not even balk at the wine bill. Nor did they blink at covering the extra two hours for whatever expenses our guests had amazed. I never doubted, as the evening went on and on, that it was worth every penny even if it would ultimately leave my pocket. It did not.

Remember when I said that Wesleyan was the best place for me to be? I was ready to cover excess cost. The evening was SO successful, and Wesleyan did not blink.

# CHAPTER 14

# MY LAST IPCC HURRAH

Sadly perhaps, the meal in Buenos Aires was the highlight of my last IPCC engagement.

I had been working myself into the ground (quite literally, I feared).

I wanted to continue with all of my obligations, but my body said "no!" - usually in the middle of the night after 3 hours of sleep. I would not listen then, but I would collapse in the late afternoon.

Then came a call from 1600 Pennsylvania Avenue. Why did I say yes when I was close to bailing on the IPCC? A good question since the call from the White House came to me on a summer's day on my deck when I had already decided that I had to pull back. I had already promised my wife that I would pull back.

Recalling my over-commitment and my commitment to my wife who was on the chaise beside me, I initially declined.

Who was that, Linda, asked aggressively having heard only one-half of the conversation. "Just the White House, I replied very proud of myself. I declined being co-Vice Chair of the National Climate Assessment."

"You cannot say 'No' to the President", she explained loud enough for the neighbors to hear. It was not exactly the President on the phone when I declined, but it was close enough. I called back within minutes to change my mind and accept.

Thinking that I could have a larger effect on US policy, I was pleased that I could retire abruptly from my IPCC responsibilities. Many of my colleagues were not amused. But they performed admirably without me.

My name is still on various chapters of the AR5 and people still remember that I invented the detection/attribution confidence matrices that permeated the entire Working Group II report. You may have noticed that I am proud of that contribution. But it was my colleagues who had completed my tasks. I am forever in their debt. Here is their evidence:

**#166** Cramer, W., Yohe, G., Auffhammer, M., Huggel, C., Molau, U., Assuncao Faus da Silvia Dias, M., Solow, A., Stone, D., and Tibig, L., "Detection and Attribution of Observed Impacts", in *Climate Change 2014: Impacts, Adaptation and Vulernability. Part A: Global and Sectoral Aspects. Contribution of Working Group II to the Fifth Assessment Report of the IPCC,* Cambridge: Cambridge University Press, 2014.

# CHAPTER 15

# THE NOBEL PEACE PRIZE

*The Lede: And then there was this. I will always remember that I should have heard the news from the TV or the Internet early in the morning. Maybe from a phone call. I heard the details from the Internet and TV, but only after the fact... I was at the computer upstairs doing ordinary early morning stuff, preparing for a day of teaching. Linda came up the stairs well before 7AM and asked if I had heard the breaking news that the IPCC had won a share of the Peace Prize for 2007. "No way!" was my response. And then we hugged, because we knew that it was true. So, really, I heard the news from Linda, with only a few minutes to celebrate.*

**Figure 14-1. Al Gore and Pachauri sharing the Prize in Oslo on December 10, 2007.** With the approval of its members at the urging of the senior members of its author teams, the IPCC used its Prize money to fund early career climate scientists from around the world.

I prepared something like a statement for the press before I got in the car to go to school. I had classes that day. Since I was senior member of the IPCC with lots of experience with people running for the Senate, I was well trained about the necessity to prepare for media. Even if nobody called, you had to be prepared.

Stuart Shlien, a very good friend, was the first to call (on my cell while I was driving near Valley View School about 30 seconds from my home). I pulled over into the Valley View Parking Lot (it was still very early and so it was completely empty). "Do I really know somebody that just won a Nobel Prize?" he asked. "Well, yes," I responded. I was not well prepared for that.

I said the same thing happened when Andy Revkin, then at the New York Times (NYT), called. I received his call almost the minute after I got to my office (still before 7AM). "Closer to the Prize than anybody I know," was his response when I explained that it was the IPCC that won the Prize.

He was right and my name was not on the Diploma. But I was a very senior member who had contributed for more than a decade to their work. In the words of the Nobel Committee that I did receive, I was awarded for "contributing to a sharper focus on the processes and decisions that appear to be necessary to protect the world's future climate and thereby reduce the threat to society".[32]

Hearing from Andy really got my attention. Hearing from the Norwegian Nobel Committee was reinforcing. Both made me feel that Gary Yohe had actually accomplished something very noteworthy; perhaps his father would have been proud and not disappointed.

I received more than 100 e-mail congratulations from 6 continents within 24 hours. The only time I received that many messages from friends who live all over the world was the day after 911. Then they were then expressing sorrow, disbelief and sympathy. This day, they were expressing the opposite. Thanks to all.

Wesleyan held a reception for me at the president's house several days later. Wow! Before I would retire 12 years later, the picture of college row that they gave me hung in my office, right next to a 4" by 6" copy of the Nobel certificate. I saw no need to dwarf the office with the real Nobel certificate, and I was proud to show Wesleyan that it was there. It was what it was.

---

[32] The Norwegian Nobel Committee, Oslo, October 12, 2007

I was still just Professor Yohe – a status that I learned from Tjalling Koopmans, James Tobin, Joseph Stiglitz, and finally confirmed by William Nordhaus in 2018.

Bill won the Economics Prize (with his name on it) in 2018. Like me, he dismissed the press for the morning, and went to class. Watching that, I recognized quickly that he, and the others in my past, had their names on Prizes and I had IPCC on mine. That was fine. When Bill won his Prize (see chapter 19), he said to me out of the clear blue sky in Snowmass: "I finally caught up to you."

# CHAPTER 16

# INSURING AGAINST RISK

*The Lede: one way of dealing with risk is through insurance. Sometimes, insurance comes in the form of hedging against a bad event by being careful to lower the likelihood of its occurrence. Other times, it comes in the form of taking action to lower the consequences of an event. In either case, insurance is a form of investment in which money is spent, now with the expectation of yielding future benefits in the form of lower cost. However, insurance is not a panacea.*

Consider, for example, coastal residents' insurance-based responses to increased flooding risks caused by human-induced sea level rise. Suppose that the residents (the demand side of a flood insurance market) think that the premium offered by the insurer and certified by a regulator is too high even though it is supported by the best available climate science—that is, they think that the quoted premium is greater than any credible estimate of the probability of loss from flooding (simple math says that the premium should be set equal to the probability).

Perhaps their subjective confidence in climate science, or in the connection between impacts and socioeconomic consequences, is low. Or, perhaps they have grown accustomed to premiums that were calculated on the basis of historical losses and not on the basis of projected future losses — losses that science now projected to become larger and more frequent in the future. They may also think that they are entitled to the subsidized system in which the Federal Emergency Management Agency (FEMA) covers any and all extreme damages at a minimal participation fee under the National Flood Insurance Program (NFIP).[33]

For any or all of these reasons, a resident could underinsure, meaning that he or she would not be paying at a sum that will be sufficient to cover actual

---

[33] https://corporate.findlaw.com/law-library/how-the-emergency-national-flood-insurance-program-works.html ?DCMP=google:ppc:K-FLPortal:10313486553:103002902536&HBX_PK=&sid=1014715&source=google~ppc

expected damage. This leaves damage to repair with no external source of funds to pay.

The question therefore arises of who should pick up the cost. Since, in conventional economic terms, rational individuals would take advantage of properly priced insurance opportunities in their own best interest, society should not pay for losses that could have been covered by individuals. This general observation, applicable beyond the specific context of insurance, can be applied in the context of climate change. It is controversial unless there is complete confidence in the science.

Of course, insurance companies need information about the likelihoods and consequences of climate-related events if they are to write policies to cover potential damages. In some cases, insurance coverage is not feasible because the risks are not known - or, if they are known, they have not been made public by the experts who have investigated them. One more significant problem is becoming widespread: the past is no longer the prologue of the future.

Recent experiences in the United States have underscored this point in a different context. The 2020 wildfires in California and the hurricanes of 2020 illustrate how extreme events are increasing in intensity and frequency, and can "pile on" to amplify one another in specific places – the damage of the collection exceeds the sum of the damages of each risk taken alone.

In California, out of the state's largest 20 fires (in acres burned), only three had occurred prior to 2000; nine of the biggest 10 had occurred in the nine years since 2012. In 2017, 9,270 fires burned a record 1.5 million acres. The Mendocino Complex fire in 2018 became the (then) largest wildfire in California history. Historic drought and unprecedented heat marked 2020, even while never before seen rain and associated flooding occurred elsewhere.[34]

The decade finished with a less noteworthy year in 2019, but then came 2020. A new largest fire in Californian history, the Complex fire, started in August 2020. Soon thereafter came the third, fourth, fifth, and sixth largest wildfires in the state's history. By October 3rd, these five conflagrations and nearly 8,000 other more "ordinary" wildfires had killed 31 people and

---

[34] https://yaleclimateconnections.org/2021/09/never-before-nb4-extreme-weather-events-and-near-misses/

burned more than four million acres. Incredibly, on that day, all five of those fires were still burning.[35]

Human actions have, of course, increased fire risk. On the consequences side of the risk calculation, catastrophic damage to life and property has increased markedly because more people have moved into vulnerable forested areas, putting their lives and property at risk and setting more inadvertent blazes. Changes in forest management have also contributed, because fire suppression policies reduced the frequency of blazes that could reduce the fuel reserves built up in forests. However, these non-climate contributors to increased fire danger have not increased sufficiently to fully account for the recent devastation.

The change in the various individual factors that create wildfire threats cannot explain the devastation if taken one at a time. Many of the 2020 fires were caused by a record number of dry lightning strikes. They were not solely the result of climate change, but they fed into a witches' brew of conditions that are all linked to global warming. The lightning strikes and other points of ignition hit in the midst of a record drought and heat wave that had lasted for weeks on end. Years of bark-beetle infestations had produced large stands of dead trees, because warmer winters had increased springtime beetle populations, and decades of gradual warming had extended the western fire season by some 75 days. Taken together, these contemporaneous influences reveal that the issue is not just what sparks the fires. The larger problem is the context within which they started, and how quickly they spread once started.

A similar story can be told about damage from tropical storms. Hurricanes Harvey in 2017 and Florence in 2018 dropped historic amounts of rain after making landfall and then stalling over Houston and North Carolina, respectively. In the summer of 2020, hurricanes Laura and Beta followed suit, causing extreme rainfall totals and substantial damage from storm surge. Their behaviors mimicked Dorian over the Bahamas in 2019. Finally, in 2017, Hurricane Mike traveled more than 100 miles inland from landfall along the Florida panhandle only to stall over Albany, Georgia, long enough

---

[35] CalFire (2021) Incidents and events – The 20 largest California wildfires, https://www.fire.ca.gov/media/4jandlhh/top20_acres.pdf

to deposit in excess of 5 feet of rain in some locations and around 4 feet across half of the state.[36]

Near-record high ocean and gulf temperatures have allowed more tropical depressions and now more non-tropical low pressure systems to develop into dangerous hurricanes. At the same time, the decrease in the summer temperature difference between the Arctic and the tropics has weakened steering currents in the atmosphere, causing storms to move more slowly. In addition, sea-level rise, one of the most obvious results of decades of rising temperatures, has compounded risks posed by storm surge.

The expanding consequences of compound fire and flood events are also having negative effects. Many of the worst fires and hurricanes have exploded so quickly and have spread so erratically that human evacuations have become "moment's notice" emergencies. As with residents of the southeastern and Gulf coasts, residents from California and Oregon have had to retreat from harm's way as quickly as possible.

New information from the ongoing scientific process can open new doors of inquiry, to be sure. More usually, though, as noted above, new evidence hardly ever implies either quickly changing conventional wisdom entirely or reversing its content.

In any case, it is critical that some protocol, like what has become standard in IPCC assessments, be followed in bringing new science into existing assessments. Only then will new reports affecting the confidence with which a piece of conventional wisdom is held have credibility.

More on this in Chapter 17 that is specifically focused on "detection *and* attribution".

---

[36] https://yaleclimateconnections.org/2020/10/multiple-extreme-climate-events-can-combine-to-produce-catastrophic-damages

# CHAPTER 17

# DETECTION AND ATTRIBUTION

*The lede: it has long been known that climate change has advanced beyond the point where mitigation alone, even with adaptation, could solve the problem. From chapters 9 and 10, recall the fundamental conclusion of the Fifth Assessment Report of the IPCC: "[r]esponding to climate change involves an iterative risk management process that includes both mitigation and adaptation and takes into account climate change damages, co-benefits, sustainability, equity, and attitudes toward risk".*

*It turns out that decision makers cannot respond to anthropogenic climate change looking into an uncertain future without being able to attribute to the changes that they detect to a specific source of stress, like climate change or population growth. It is impossible to project ranges of future climate driven impacts and their consequences without an understanding the correlation between markers of human induced climate change (e.g., global mean temperature reflected in local or regional temperatures) and the detected impacts.*

This chapter descends into some technical weeds, but they are critically important weeds. Understanding the definition and application of the statistical definitions of "attribution," "prediction," and "projection" is essential. "Prediction" denotes model-derived estimated values for an output variable given a vector of input values, usually selected from within (or close to) the domain of detected data (the observations). "Projection" denotes estimates of the output variable from input values that are expected to lie outside their observed domains based on confidence in our understanding of underlying processes. Confidence there depends on the strength of "attributing" observed outputs to climate variability, and perhaps to trending anthropogenic climate change as compared to other confounding factors like population growth or economic development.

My first foray into detection and attribution came in a joint paper with Camille Parmesan, published in *Nature* in 2003 (#60). It was born at a Working Group II authors' meeting in Lisbon (see Chapter 13). In that paper, we worked on distinguishing the causes of detected changes in the ranges of butterflies between anthropogenic climate change (warming) and other stressors like local pollution, population, or new development patterns. The data were nearly 300 papers in the published literature from authors who tried to make that distinction in a particular location somewhere in the world. The question was – does the collection of these localized attribution studies (to local climate change or other drivers from whatever source) support high confidence in a conclusion that we have discovered a ***globally coherent signal*** of the influence of anthropogenic climate change?

Confounding factors were a problem, so our novel statistical construction could produce only medium confidence without some help from another source. The other source?

Camille had observed that ranges of Edith's Checkerspot butterfly (a widely studied butterfly) typically moved north, or up mountains. That meant that, while populations were declining along the southern boundary, which was warming beyond the ideal, they were expanding along the northern boundary, where warming was moving toward levels with which they were more comfortable.

Only anthropogenic climate change could switch signs like that; and since a large fraction of our data set detected this duality in their data, we could finally report ***with very high confidence*** that we had detected a global signal.

Why? Because we were certain, in those studies, that the authors had correctly attributed all of the detected range change to human induced global climate change (nothing else can switch signs).

As an aside, this paper is my most cited referenced piece of work, with more 12,500 citations through the summer of 2022 and counting (not bad for a paper published in 2003).

The IPCC conclusion about iterative risk management mentioned in the lede makes it clear that investments in adaptation (or mitigation, for that matter) depend upon efficiently processing information about the magnitudes of the consequences of observed and projected climate change as well as descriptions of the relative likelihoods of both – characteristics that will have been detected and quantified from historical data and then, perhaps, attributed to climate change and its anthropogenic sources so that ranges of future projections can be authored.

From the perspective of real time reactive adaptation, simply detecting changes that may have been driven by local climate change and/or other factors may be sufficient to inform effective responsive decisions for the short-run. Information required to assess decisions about anticipatory adaptation, as well as long-term development projects, is always much more complicated. By their very nature, they rely on the attribution of detected changes to human sources of climate change that can be quantified statistically and differentiated rigorously from the effects of other confounding factors that are typically driven, predictably, by underlying factors, like the pace of economic growth correlated with the pace of population growth.

Confidence in attribution and its quantitative calibration is, therefore, critical to efforts designed to project ranges of possible risks that adaptation and investment decision-makers do, or at least should, take into account. Chapter 18 of the contribution of Working Group II to the IPCC AR5 (#166) reported the possibility of assigning greater confidence to the projection of climate change related phenomena than to the detection and attribution of changes that have already been observed. Their Figure 18.2 is replicated here as Figure 16-1.

Personal note: Accepting this the figure as a means to frame detection and attribution in your IPCC chapter was the ticket to a free dinner in Buenos Aires (see Chapter 13).

Chapter 18 authors also spent enormous effort with regard to detection and attribution with regard to Reasons for Concern (see chapter 6); the abstract below highlights their conclusions.

Working from these conclusions, several fundamental questions emerged before my eyes:

- How can the confidence in projected vulnerabilities and impacts be greater than the confidence in attributing what has heretofore been observed in ways that are consistent with expectations derived from first principles of statistical analysis?
- Are there characteristics of recent historical data series that do or do not portend achieving high confidence in attribution to climate change in support of framing adaptation decisions for some point in an uncertain future?

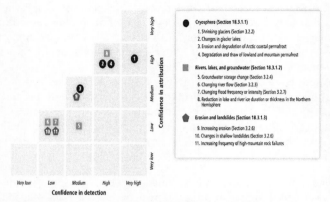

**Figure 16-1. Correlations between confidence in detection and confidence amplified by process understanding in attribution.** Source: Figure 18-2 in WGII AR5.

- What can analysis of confidence in attribution tell us about ranges of "not-implausible" extreme futures (that are found in the tails of the distributions of impacts) vis-a-vis a static (but stochastic) future assumed from a static climate system?

Answering these questions in an adaptation context is essential because of the long-term nature of some adaptations, as well as for plans to reduce greenhouse gas emissions. All three answers require an understanding of the underlying physical and social processes by which confidence in impact projections can legitimately be evaluated to illuminate the foundations of strategies for iterative risk management. That is to say, this understanding is necessary if the science and subsequent defense of adaptive and mitigative response decisions are to be able to navigate what might otherwise be viewed as both a contradiction of statistical rigor and an obstacle for rigorous policy evaluation of adaptation strategies.

These are the answers which can explain why the science can support attribution of the recent drought in Texas (from 2011 through the end of 2013) to anthropogenic warming, while it cannot yet support a similar conclusion for the recent five year California drought that also began in 2011 – a location where diverse topography and the proximity of an ocean confound the statistics of what would seem, at first blush, to be a "no-brainer" attribution.

The key to the answer to all three questions is a reasonable expectation that, based on a growing number of observations indicating the unequivocal anthropogenic drivers of the observed climate warming trend, researchers should expect that the impact of more micro scale climate changes, and consequentially their micro scale manifestation of a globally coherent "climate signal," will they

- increase substantially with time, while
- the impact of confounding variables like societal or geographical characteristics could remain constant (or at least trend less significantly in line with observable and well-known driving socio-economic variables).

Borrowing significant amounts of language from #214, referenced below, recent experiences with wildfires are perfect examples of this phenomenon. Only three of the state's largest 20 fires (in terms of acres burned) had burned prior to 2000, but nine of the biggest 10 have occurred since 2012. That is, extreme events were becoming more likely - and they are growing larger, too. In 2017, 9,270 fires burned a record 1.5 million acres. The Mendocino Complex fire the next year became the largest wildfire in Californian history. And then came 2020.

A new largest fire in Californian history, the Complex fire, started in August of 2020. Soon after, the 3rd, 4th, 5th, and 6th largest in history were ignited. By October 3rd, these five conflagrations had combined with nearly 8,000 other, more "ordinary" fires to kill 31 people and burn more than four-million acres, and, on that day, all five of those fires were still burning. Simultaneously! That is like having 5 "500 year" floods raging at the same time in the same state. But for climate change, that is impossible.

How so? Wildfire has certainly been a natural part of the forest environment for a long time. However, by the early 1950s, wildfires were causing sufficient damage with sufficient frequency to provoke efforts to reduce what was seen as their main cause – human behavior. "Only you can prevent forest fires" was the mantra of the times, and Smokey the Bear was the mouthpiece.

Only a few decades later, however, changes in the climate had begun to contribute to increased fire risk. More intense droughts played a role in some years, as did extra strong heat waves. Also, milder winter temperatures were fostering the expansion of a major forest pest, the Pine Bark Beetle, which was killing large areas of forest and thereby further increasing the supply of fuel.

Of course, in 2020, part of the increased fire risk could still be attributed to human actions. Damage to life and property had increased markedly as more people moved into vulnerable forested areas. It is obvious that more people in the woods meant more inadvertent blazes - especially stupidly designed gender-reveal events for expectant mothers that sent pink or blue fireworks into the forest. Changes in forest management contributed, too, because fire suppression policies on federal land reduced the brush-clearing value of deliberately set control blazes (sometimes known as "good fires").

These non-climate causes of increased fire danger had increased with sufficient speed over the decades to account for the devastation of the last few years in the time series. There was more to explain, however, and it came down to understanding how, in responding to rising global temperatures, nature can produce "2-fer" or even "3-fer" combinations of influences on local environmental conditions.

In California in 2020, those compounding climate effects created a witches' brew of conditions that are all linked to global warming:

- Exaggerated numbers of dry lightning strikes in the midst of a record drought;
- record heat for days at end in July and August;
- exploding infestations of bark beetles producing large stands of dead trees; and
- decades of gradual warming that had extending the western fire season by some 75 days.

Taken together, these contemporaneous impacts make it clear that the issue is not just what sparks the fires. The larger problem is the context in which they start, and how quickly they spread once started, especially when several intensifying influences are also present. For climate change, though, they would not be so severe.

So, what are the answers the three questions above? For the first, including robust understanding of underlying processes that describe output risk along projected ranges of how the future might unfold can, in some cases, uncover bifurcations between the distributions of outcomes calculated with and without anthropogenic drivers. As a result, forward-looking adaptation can be well-informed of what the future might bring. These bifurcations are likely to emerge along high climate change scenarios (so they may be delayed by effective mitigation); in other words, we cannot forget that mitigation matters. Conversely, bifurcations may be obscured by confounding factors, especially if distributions of climate change uncertainty grow slowly as the future unfolds.

And the second question? The characteristics just noted show the answer, here, is positive. Bifurcation can add urgency to adaptation decisions that might otherwise be questionably appropriate when only current climate variability can describe the decision-making environment.

And the third? Looking for bifurcations means focusing attention on the tails of impacts distributions. By their very construction, bifurcations occur when the high 95th percentile future assuming no climate change deviates measurably from the 5th percentile future assuming that anthropogenic climate change is driving the future. Put another way, the worst possible future becomes the best possible future – a concerning but not unmanageable conclusion for long-term adaptation planning. It follows that anticipated dates of bifurcation assume even more significance, because they can signal a discontinuous change in the decision environment.

The takeaway from all of all of this is still that determining whether a given magnitude of output risk can be attributed to a high climate signal or a high confounding factors baseline can be a daunting challenge. This is particularly true with respect to carefully distinguished cases where responses to climate induced risks can only be informed by predictions of growing collections of current observations, rather than effectively extended data sets that include credible ranges of future projections.

# References, abstracts and links connected to my January 2023 CV

**#60** Parmesan, C. and Yohe, G., "A Globally Coherent Fingerprint of Climate Change Impacts across Natural Systems", *Nature* 421, 37-42, January 2, 2003.
Causal attribution of recent biological trends to climate change is complicated because non-climatic influences dominate local,short-term biological changes. Any underlying signal from climate change is likely to be revealed by analyses that seek systematic trends across diverse species and geographic regions; however, debates within the Intergovernmental Panel on Climate Change (IPCC) reveal several definitions of a 'systematic trend'. Here, we explore these differences, apply diverse analyses to more than1,700 species, and show that recent biological trends match climate change predictions. Global meta-analyses documented significant range shifts averaging 6.1 km per decade towards the poles (or meters per decade upward), and significant mean advancement of spring events by 2.3 days per decade. We define a diagnostic fingerprint of temporal and spatial 'sign-switching' responses uniquely predicted by twentieth century climate trends. Among appropriate long-

term/large-scale/multi-species datasets, this diagnostic fingerprint was found for 279 species. This suite of analyses generates 'very high confidence' (as laid down by the IPCC) that climate change is already affecting living systems.

**#166** Cramer, W., Yohe, G., Auffhammer, M., Huggel, C., Molau, U., Assuncao Faus da Silvia Dias, M., Solow, A., Stone, D., and Tibig, L., "Detection and Attribution of Observed Impacts", in *Climate Change 2014: Impacts, Adaptation and Vulernability. Part A: Global and Sectoral Aspects. Contribution of Working Group II to the Fifth Assessment Report of the IPCC,* Cambridge: Cambridge University Press, 2014.

- Evaluation of observed impacts of climate change supports risk assessment of climate change for four of the "Reasons for Concern" developed by earlier IPCC assessments. Impacts related to Risks to Unique and Threatened Systems are now manifested for several systems (Arctic, glaciers on all continents, warm-water coral systems).
- High-temperature spells have impacted one system with high confidence (coral reefs), indicating Risks Associated with Extreme Weather Events. Elsewhere, extreme events have caused increasing impacts and economic losses, but there is only low confidence in attribution to climate change for these.
- Though impacts of climate change have now been documented globally with unprecedented coverage, observations are still insufficient to address the spatial or social disparities underlying the Risks Associated with the Distribution of Impacts.
- Risks Associated with Aggregated Impacts: large-scale impacts, indicated by unified metrics, have been found for the cryosphere (ice volume, high confidence), terrestrial ecosystems (net productivity, carbon stocks, medium-high confidence), and human systems (crop yields, disaster losses, low-medium confidence).
- Risks Associated with Large-Scale Singular Events: impacts that demonstrate irreversible shifts with significant feedback potential in the Earth system have yet to be observed, but there is now robust evidence of early warning signals in observed impacts of climate change that indicate climate-driven large-scale regime shifts for the Arctic region and the tropical coral reef systems.

**#214** Yohe, G., Jacoby, H., and Richels, R., "Multiple extreme climate events can combine to produce catastrophic damages", *Yale Climate Connections,* October 9, 2020,
https://yaleclimateconnections.org/2020/10/multiple-extreme-climate-events-can-combine-to-produce-catastrophic-damages/
Concurrent extreme climate events can amount to a challenging 'two-fer' or even a 'three-fer' or 'four-fer' in terms of adverse impacts. The result are consequences beyond what might have been expected taking on impact vector at a time and adding them up.

# CHAPTER 18

# ENGAGING IN THE PUBLIC DISCOURSE

*The lede: it turns out that communicating climate science and climate policy is really hard. It takes lots of concentration, discipline, and attention to detail; otherwise, it is easy to look or sound foolish or self-important. There are frustrations over inappropriate homage to the mantra of "fair and balanced" coverage by media outlets that feel obliged to tell both sides of a story, even when the consensus is more than 80% on one side.*

In the Federal Register, it always looks like the climate debate is at best a 50-50 proposition because both sides usually get to choose the same number of witnesses. There are frustrations over alternative facts created without any evidence by those who dismiss climate risks. They want to sow doubt and uncertainty in the minds of the public. For example, if climate change actually exists, it is caused by sunspots, or population growth, or the misalignment of the earth's orbit, and so on. NOT.

Steve Schneider was the only one who was immune to the veracity assault in one-on-one debates because he read everything, remembered everything, and could cite chapter and verse. But that was not enough. He could never make a mistake. And he never did.

In my work with fellow researchers on the third National Climate Assessment (NCA3), assessments of the Intergovernmental Panel on Climate Change (IPCC), the New York (City) Panel on Climate Change (NPCC), various panels and committees of the National Academy of Sciences (NAS), other government committees from within the National Oceanic and Atmospheric Administration (NOAA), the National Aeronautics and Space Administration (NASA), the Environmental Protection Agency (EPA), the Department of Commerce, etc, we always scared to death. We worked very hard never to publish "slow moving targets" that would include an error.

The aspirational goal was "no mistakes anywhere in the text". That did not mean that we didn't report on low probability but high consequence risks;

it just meant that we were honest describing the science that supported assessments of both components of risk.

We never understood why the dismissive side did not have to play by the same rules. The Heartland Institute or Fred Singer could cherry-pick data or studies, or just make things up, and get away with it. Jon Christy kept citing one of his papers in his public talks for years after he was forced to publish a retraction.

Bjorn Lomborg is an exception. He is not dismissive, but he does worship at the holy grail of cost-benefit analysis in his Copenhagen Consensus exercise and so he does not think that climate change is particularly important.

I contributed three times to his exercises, writing two position papers, four years apart, in support of the problem of climate change, and discussing the process with John Bolton (not a Nobel Laurette) among others in between. Mr. Bolton agreed that climate change was a problem, but he did not think that it should be a priority for a world facing multiple critical stresses and problems with an artificial budget constraints of $X billion in one year (he got to set the limits).

With the first paper, written with Richard Tol, we finished last in the rankings of global problems prepared by Bjorn's evaluation committee that he populated with Nobel Laureates.

I later criticized Bjorn for his characterization of our work in the *Guardian*. I wrote a critical opinion piece for the same venue which highlighted what I thought was his mischaracterization (I still do). It caused Bjorn trouble at home, and so he responded. We could have gone back and forth for quite a while, but we spoke on the phone and decided to write a joint piece. A week later, the *Guardian* published something called "It is not about us" where we both agreed that the climate was changing and that humans were largely to blame (#102). Choosing the right response was just a matter of degree – especially since a few billion dollars could really help reduce the incidence of, for example, malaria in Africa. We have been sort-of-friends ever since.

One episode at the end of the NCA3 process shows how seriously we took the aspirations that we never commit a public mistake.

We had divided the country into eight regions on the North American continent and Hawaii, and we had prepared two-page overview documents for climate and impacts for each of those regions – as diverse as they were.

Two or three weeks before the June release of the NCA3 in 2014, John Podesta (then President Obama's Chief of Staff) got the idea of publishing two-page summaries for each of the 50 states derived from our eight-page chapters of regions that covered up to 10 states. We had traceable accounts wherein the authors could indicate where, within each region, this or that manifestation of climate change had been detected and attributed.

Mr Podesta wanted to assign White House interns to specific states so that they could cover the effect of climate change across the country one state at a time.

Kansas was the trial balloon for his idea. Repeated drafts applied climate change descriptions for the entire Midwest region to Kansas. The trouble was that the Midwest region extended from Minnesota to Texas, so most of the climate descriptions included in the draft Kansas summaries were misplaced and so completely wrong. Repeated attempts to fix the problem all failed, but Mr. Podesta persisted right up to the June 6th release of the NCA3 (https://www.nca2014.globalchange.gov) in the Rose Garden.

Jerry Melillo, T.C. Richmond, and I (as the leadership troika for the three-year NCA3 project) finally had to make a stand. We sent word to the White House on June 5th through John Holdren that their releasing the hopelessly inaccurate state summaries would cause us to take our names off the Assessment. We told the President through John that we fully understood that we, and not the release of the Assessment, would be tomorrow's news story across the country if we did that.

It took a New York minute for the White House to blink; word from a non-linear office sufficed. The state summaries were not released.

We could have satisfied Mr. Podesta's plan if we had started 18 months earlier, but 3 weeks and limited coverage at the requisite level of detail were simply inadequate.

Living with dismissive mouthpieces on the planet means that I had to be prepared to speak with one or many. Few of them are open to changing their minds in the face of scientific facts. They are more interested in making audiences unsure of what is true. You have to be prepared with a rigorous and accessible retort and a library of references. So, when they say

"climate change is a hoax!",

I must say:

"have you ever taken a look at the temperature record? Every decade since 1950 has been warmer than anything in its past five years. Twelve of the hottest 15 years on record have occurred during the last 15 years. 2022 might not have been a record, but it was in the top three, and it was hotter than anything that Europe had EVER seen.

Then, they say:

"Well, the climate may be changing, but that causes little harm."

So, you must say:

"Do you ever watch TV? The intensity and frequency of extreme weather events that have been attributed to human induced climate change (droughts, flash floods, riverine flooding events, extreme heat waves, coastal storms, cyclones and hurricanes, severe cold snaps, wildfires, etc..) have exploded over the past 4 years, accounting for more than 50% of the $1.5 trillion in disaster losses suffered across the United States since 1980.

Coverage of these events is on the TV every night, from home and abroad; it demonstrates that loss of life is a real and frequently unnecessary if you stay well-informed." TV weather correspondents show maps of storms that span a continent – week after week after week with heretofore unknown atmospheric rivers and frigid bomb cyclones. Never before has this happened.

They respond:

"Even if damages are occurring, there is nothing that we can do about it",

So you must say:

"reducing emissions of heat-trapping gases can slow the pace of climate change" That is unequivocal. Careful forward-looking investment in adaptation can save lives and treasure, but it cannot eliminate either"

Recently, the conversation has continued through one more iteration: "Even if there is something that we could have be done, it would have been too late. We should just give up and enjoy the ride without worrying,"

Then you say that this is not playing "Thelma and Louise" with the planet. Instead, you cite Jim Valvano:

"Don't give up! Don't ever give up!"

It worries me that the dismissive side has crafted one sentence statements that are easily understood even if they are false. I recognize that I have to work to express complicated stuff that takes mind-numbing paragraphs to explain.

My colleagues and I (Richard Richels from the Electric Power Research Institute, Henry (Jake) Jacoby from MIT, and Benjamin Santer late from UCLA) have been writing opinion and fact pieces since the fall of 2019. We will have published a collection of our essays through Springer Nature in the spring of 2023; Part IV entitled the "Yale project for the campaign season" was commissioned by Yale Climate Connections for the weeks leading up to the 2020 election.[37] It begins with our collective elaborations of what we would like to hear from candidates for public office, but it continues with a series of articles that provided the details of our responses to the dismissive arguments catalogued above – contradicted by the volume below.

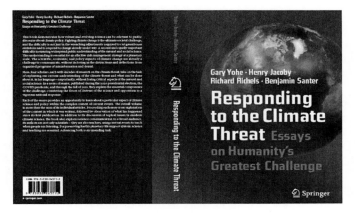

**Figure 18-1. Cover of the SpringerNature collection**.

.

---

[37] Part IV in Yohe, G., Jacoby, H., Richels, R., and Santer, B., 2022, *Responding to the Climate Threat – Essays on Humanity's Greatest Challenge,* The Netherlands: SpringerNature.

# CHAPTER 19

# THE NOBEL PRIZE AGAIN

*The lede: Professor William Nordhaus won the 2018 Nobel Prize in Economics on October 8, 2018. That news was exciting enough, because it validated both his and my own spending of much of our lives working on climate change and framing climate. But then it trickled down to me.*

On October 15th, Bill invited me to be one of his four non-family guests at the Awards Ceremonies and Events in Stockholm.

Can I go? I was still teaching and commuting between Connecticut and Maryland. The unanimous opinion from family was that I had to go.

His invitation indicated dates (December 8th through the 10<sup>th</sup>) that were most important). I made sure that I was available to travel abroad spanning those dates but they were not constraints. It would be my first time abroad in 2 years, and I did not want to miss anything. Since I was going, I would be there for all seven days.

To be precise and very formal, William D. Nordhaus had been awarded the Sveriges Riksbank Prize in Economic Sciences in Memory of Alfred Nobel for 2018 – otherwise known as the 2018 Nobel Prize in Economics. The attribution justifying his award was amazingly concise: for "integrating climate change into long-run macroeconomic analysis". Bill shared the Prize with Paul Romer from NYU, who was selected for the equally concise reason: "integrating technological innovation into long-run macroeconomic analysis".

That was nice symmetry that did not go unnoticed in their Nobel Lectures, but that is a story for below. For the record, here is the longer description of Bill's work that caught the Nobel Committee's attention:

> Nordhaus' findings deal with interactions between society and nature. Nordhaus decided to work on this topic in the 1970s, as scientists had become increasingly worried about the combustion of fossil fuel resulting in a warmer climate. In the mid-1990s, he became the first person to create an integrated assessment model, i.e. a quantitative model that describes the

global interplay between the economy and the climate. His model integrates theories and empirical results from physics, chemistry and economics. Nordhaus' model is now widely spread and is used to simulate how the economy and the climate co-evolve. It is used to examine the consequences of climate policy interventions, for example carbon taxes.

The welcoming note in my room.

Bill and myself on the morning of Day #1.

Bill & Romer after their lectures.

Most of the Climate Club.

Top of the mountain.

**Figure 18-1 Pictures in an exhibition from Stockholm**

All of this is true. I have used DICE and RICE (his two ever-evolving models) to this day to support some of my work. Wh invent the wheel?

I have always used the insights that he derived from his modeling in all that I do – to frame analyses but more importantly to reveal fundamental questions. It is not about the numbers, he taught me; it is about the insights and the underlying processes that allow us to push science forward.

Per that insight, Bill and I have come to coexist on the interface of two approaches to the climate problem. His grows from dynamic cost-benefit maximization, which he now uses to explore alternative approaches (like shooting to hold warming below 2 degrees Centigrade to maintain tolerable risk profiles at minimum cost) and Carbon Clubs.

Mine works from an iterative risk-management perspective calibrated in many different metrics (like currency, or human lives or...) that always keeps track of the Bill's "optimal" solution for grounding.

We can work together.

On October 8th, the day of the announcement of his award but before noon, I sent a Tiffany pen to Bill in recognition of his being awarded the Prize. It arrived at Humphrey Street the next day to the puzzlement of Bill and his wife Barbara (I found out later).

I had been inspired by the scene in "A Beautiful Mind" where, just after winning the same Prize, John Nash sat down at a table in the Princeton Faculty Club. Many of his colleagues approached him at his table and gave him their congratulations that were embodied in their personal pens. Barbara thought that I knew that Bill is fascinated by pens and that I wanted to contribute to his collection. I did not, but that made it all the more special.

My invitation to Stockholm arrived on October 15th of 2018:

Dear Gary,

I have the opportunity to invite a handful of guests to the ceremonies in Stockholm in December. The key days are December 8 through 10. I would love to have you for being a central participant in work at the National Academy and for a continuous distinguished work on the integrated assessment of climate change. If you can join us, I will send more details on the plans, dates, and ceremonies.

Best, Bill.

Once Barbara and I began to communicate about my impending trip to be part of the 14 person Nordhaus tribe, that would name itself the "Stockholm 2018 Climate Club", I was able to explain my motivation for the gift. She let Bill know.

The dates conflicted with a long-standing professional trip that Mari had on her calendar. Linda and I had agreed that we would take care of the twins in her absence. I was torn, but everyone agreed that I had to go, because this opportunity was not to be missed. So, I began to prepare. They would manage in my absence, but perhaps not without some cost.

My passport was the next source of concern. I had sent my expired passport to the government to be renewed a few weeks earlier, but I had not asked for expedited service. Now, suddenly, I needed expedited service to make hotel and plane reservations. I called Senator Murphy's office for help. The Senator and I had some history, not to mention Courtney and Todd who ran his first campaign for the State Legislature.

All jumped at the prospect making it possible for me to get to the Nobel ceremonies. The Senator sent me a privacy form to complete, and then went to work expediting. Their contact in the Atlanta office was also impressed with the reason for the trip, found my application in backlog, processed it, and put my new passport in Federal Express in about 3 hours. Within 24 hours of my request for help, my new passport arrived at the door of our apartment in Rockville, MD.

Unbelievable. Thank you, Senator. I owe you one.

The Nobel Committee soon sent 12 pages of instructions, schedules, another information for Laureates' guests. Since I had a new passport number in hand, I quickly followed their advice about using SAS for the transatlantic flight. I also decided that it was essential that I stay at the Grand Hotel in Stockholm – the headquarters hotel for the Prize infrastructure that was equipped with a "Nobel Desk" to distribute tickets and to handle all problems for any of us.

Then there was the requirement that I wear "white ties and tails" to the Award Ceremony and the Nobel Banquet. I contacted the recommended store in Stockholm, sent in my measurements, and rested assured that I could pick up the requisite apparel upon arrival.

After all of that, I received word that I would be greeted with VIP service at the airport; I would be met as I disembarked from the plane and transported directly to a private office. I would not have to stand in line for customs, claim baggage, or arrange transportation to the hotel – both coming and going. They would take care of everything. That was very cool. And so, the trip.

## December 6th

I left for Stockholm after my last Econ 310 class of the semester on the morning of December 6th. I arrived I Stockholm the next morning in the dark, exhausted from a busy week and a long flight. VIP service took care of me.

## December 7th

I was delivered to the hotel, but my room was not available. So, I took a cab to the tux place, had my tails and trousers fitted, and otherwise filled in the morning by trying to stay awake; I was not feeling well. \

My room was available around noon, so I got to nap before dressing for an evening initial reception with the Nordhaus group, as well as Professor Romer, and a subsequent dinner (all thankfully in the hotel). I went to bed happy about the prospect of a good rest and a good breakfast before the festival of events coming the following day.

## December 8th

Most of the Nordhaus group went on a guided tour of the Royal Palace and Old Town Stockholm in the morning.

2 PM: Nobel Lectures in Economics – Bill and Paul Romer spoke for 30 minutes each, outlining in more detail the work that had been celebrated in the Prize.

Walking out, Annabel (a grand-daughter) asked what time it was. 3:30 (PM) was my answer. She was incredulous because it was already completely dark outside. I asked her why that might be. She told me that she did not know, so I asked her to think about it. She thought about it and told me why the next morning. Spot on, she was.

7 PM: Nobel Prize Concert – The Stockholm Philharmonic Orchestra and soloist Lisa Batiashvili performed a violin concerto as well as Tchaikovsky's

4th Symphony under the baton of conductor Karina Canellakis. The performance was spectacular.

## December 9th

We had the morning off, so Jesse, Naki, and I went to the Nobel Museum instead of taking the programmed boat ride. We found the 2007 Peace Prize description, and explored many exhibits; the dominant displays involved Martin Luther King, Jr, but we also found the chair that Bill had signed.

4 PM: We all attended a reception at the U.S. Residence. The United States had no ambassador thanks to…er….., but the Charge d'Affaires hosted more than 100 guests. I met Frances Arnold, mistaking her for staff, but came to understand her work later by conversing with my daughter Marielle. Frances was very gracious and Mari was very impressed.

I learned the Nordhaus version of the rules for "Crow's Nest" and "Sock Wars" from Alex and Margo while sitting on the floor of the dining room of the Ambassador's house. "Practicing to be 'papa'!" is what I told everybody when they walked by; in reality, I had already been adopted by the twins.

6 PM: Reception for all at the Nordic Museum. I was walking from my room when Bill emerged from the elevator with Alex and Margo. "Here," he said. "Take them down to the lobby and find somebody to get them off to the reception." I was in charge of the grands…. What could go wrong? Nothing because they already liked me.

So off we went into the chaos of 3 busloads of people in the lobby. Some family arrived, but I understood that I was in charge of the pair of grands until and perhaps after we arrived at the reception. They did too, and so we had fun.

The enormous room for the reception was jammed with hundreds of people. There was no place to walk. There was no comfortable room for children except for some space under a giant Christmas tree at one end of the enormous hall where there were very few people.

Ignoring the crowds, the twins and I went down to the tree and proceeded to try to count the ornaments on the tree. They (remember that they are maybe 8 years old) agreed to divide the tree into sections. I do not remember whose idea that was, but we counted one section, multiplied by 4, and accounted for fewer ornaments on the back side of the tree. Their total was 3005 ornaments. A little over precise, …..,

Eventually the family collected under the tree so that we could go to dinner. The girls told Grandpa about the counting exercise that I had concocted, so he took a shot. He did the geometry correctly and came up with about 1500-1600.

There was an interesting debate, but the Nobel Laureate won. The children had assumed a cylinder and not a cone. My bad, but too much math.

7:30 PM: The Nordhaus family hosted a Swedish Christmas buffet at Operakallaren for the Climate Club. We got there late, and did not leave until well after midnight – eating and drinking through 12 courses of exquisite food.

This was a spectacular time at a very long table. A parallel table was occupied by a collection of dentists. They gave toasts, and we returned the favor. When it was time for our toasts, I went second; my words are appended below.

Around 1:30 AM, we left the restaurant. Lint Barrage was staying at a different hotel, so we all walked her to her door. Then we walked back to our hotel, arriving after 2AM.

Something about a Nobel Laureate walking an invited guest back to her hotel in the dark of night after midnight, accompanied by his family and other guests, just sticks in my memory. Would be a great scene for a movie. It was certainly a sign of enormous character, I think. When we were making a lot of noise near an apartment building, we all warned each other to be quiet lest we be arrested. Bill responded "you all are on your own. They are not going to arrest me."

Anyway, it was at this dinner that the Stockholm 2018 Climate Club was formed.

# December 10th

Another morning off, but a good thing. All us woke up late and barely made breakfast. I took a walk and took some pictures, but then began to worry about getting into my tails. They had been delivered to my room on the 7th, but they were still in the suit bag.

We had to be on the bus at 2:50, so it was time to put it on. I started fussing around 1 PM. Shirt sleeves were easy, and I had cuff links from Mari to wear. The studs and the collar anchor were challenging. Trousers were easy, and so were the suspenders. But the bow tie... argh!

I found it impossible to hook the tie properly while looking in a mirror. After 15 minutes of frustration and growing anxiety, I went downstairs to the lobby where the entire staff was in tails. I asked for help from the rental store, but there was no need. The concierge took me aside, adjusted my tie, and got it on my neck properly in about 30 seconds.

I went back to my room to finish dressing, and made the bus with 5 minutes to spare.

4:30 PM: The Awards Hall was spectacular. Our seats were perfect (second row mezzanine in the center). The stage was filled with members of the Swedish Academy of Science on the right and earlier Laureates on the left. The Royal family sat up front on the right, and this year's class sat in front of the Laureates on the left.

When they came on stage to trumpets and fanfare, I looked at Link and said "I think he will be nervous, now." Bill had taken the entire week in stride up until this. When he sat down, second chair from the right, his feet could not stay still and he could not figure out what to do with his hands. I was right, but not for long. When it was his turn, he did the choreography perfectly, justifiably rightfully proud – if only for a minute. This was the rightful acknowledgement of the top of the mountain.

The Stockholm Philharmonic and an opera singer performed in between each discipline's awards – physics, chemistry, medicine, and economic science. The singer's name is Christina Nilsson; she sang beautifully. She would also sit directly to my right during the Banquet to come; I found out that this was the fifth time that she had sung for the King in 2018.

7:00 PM: It was snowing a little when we left the Awards venue for City Hall and the Nobel Banquet for nearly 1600 people. The meal was planned for 3 courses and 3 hours and 45 minutes. Entertainment emerged from all sides in between courses. As already noted, Christina Nilsson sat to my right. Valentina Bosetti sat to my left; she was a colleague of Bill's at IIASA. Naki sat across, and Jesse sat to the left of Valentina. Lint sat directly behind me at the next table, so we were all close.

Our group was about 10 feet away from the King of Sweden. Across and to the left sat John Arne Hassler, a professor in Stockholm and chair of the committee that awards the Economics Prize. We started talking shop about the social cost of carbon and what happens when net emissions fall to zero. Just like the IPCC days – pay attention at dinner.

The banquet adjourned around 10:45 only to reconvene for dancing and more food upstairs in a room that was as big as the banquet hall. The room was in motion until 2 AM. Another party followed, but Lint, Jesse, Naki and I went back to the hotel for some rest. We all had to travel tomorrow.

When we got back to the hotel, we picked up our "Climate Club" hats at the front desk – courtesy of Monica Nordhaus and named for the location and Bill's latest publications about a "climate club". All game theory and the like, but the hat feels great.

## December 11th

I awoke in time for breakfast, saw Bill in the lobby one last time for a hug, and headed off for my plane. First, to the VIP building with its own security belt, customs officer, and people to check bags. I was told to wait until the last minute. Then, I was driven to the gate on the tarmac, went up the back stairs to the jetway, and walked onto the plane – last among all passengers to board the plane. All looked at me to try to decide who I was. I knew who I was: I was Bill's friend.

I slept most of the way to Newark, and then most of the way in the car back to Portland.

Traffic was awful on the way to home, but I really did not care. I was on top of the world!

# CHAPTER 20

# MY TIME WITH STEPHEN SCHNEIDER (1945-2010)

*The Lede: Steve was the general in the climate wars because he was the exemplar of all that we all wanted to be. For reference, please see https://stephenschneider.stanford.edu/References/Biography.html).*

Why was Steve a general in the climate wars? Because he knew everything, new and old. He took personal risks (and it killed him). He took on anybody on stage or on camera, and he was the only person alive who could convincingly respond in a debate to any point from any dismissive opponent by saying, "you are making that up, it is wrong, and here is why."

How could he do that? Because of his creation of and editing of *Climatic Change* which attracted cutting edge research, but also from his careful attention to the other literature. I mean, "everything". You could not surprise him with "I saw this in journal XYZ – especially *Nature* or *Science* so what do you think". He will have already seen it and he would certainly already have an opinion.

More importantly, he also knew that whatever he said in public or in print had to be true because he was never allowed to make a mistake. If it came out of Steve's mouth, it had already passed peer review. And if he were wrong, then he knew that all peer review of climate science would suspect.

At the same time, Steve never took the median or the mean for the only answer to any question about what might happen or what any of us should do – not from those of us who were worrying about climate change, and certainly any of us who were listening not from our doctors.

He was the "patient from hell", but he survived his first bout with a serious condition because he always knew more about the risks that he faced than his doctors.

I had remembered his book, and so I remembered clearly sitting under the tent in Snowmass when a phone range. Terry came up to him, whispered in his ear, and off they went (chapter 11). He picked up his stuff.

He and Terry left Colorado immediately to go back to Stanford to check into the hospital. The tests had come back, and the results were not good. Not many noticed his departure, but I had. When I asked, the front desk confirmed that he and Terry had checked out to go home.

Once it became common knowledge, we all knew that this was not good. But he recovered this time.

I summarized this story in my comments at his memorial service. I drew from my memories of the Scoping Meeting for AR5 that I included in in the Venice section of Chapter 13:

> I had finally passed the entrance exam to his inner circle. I got up in plenary at a Scoping Meeting for the AR5 in Venice and said to the Chair of Working Group I – "What you just said (that one peer reviewed paper with a contrarian conclusion would not be assessed or included in the assessment of the AR5) made the hairs on the back of my neck stand on end. We had been working in a risk management world for six years, and so dismissing a not-implausible conclusion from the peer reviewed literature with high consequences was "full of shit".

Steve defended me in Venice because, "since the Valencia plenary in 2007, our IPCC clients had said that they wanted to hear about the full range of possible futures. Our clients thereby made it clear that they want to be informed about the dark (or benign) tails of 'not-implausibility' futures."

"What Gary was trying to say was......", he had begun. Steve was very measured and very polite in summarizing my thoughts until he got to the "full of shit" part. Then he agreed. I knew, then, that I had finally passed my post-doctoral exam. Steve was pleased enough with me to defend me in front of 170+ country scientists.

We lost that battle with WGI, but I can report here that I had a smile on my face after Steve had made a two-fingered intervention from across the room (signaling that "my comments were germane to what was just said), and Pachauri (who had the microphone) had acknowledged that.

Sometime later, I heard that Steve had died while I was having breakfast at the River Inn in anticipation of spending the day attending a National Academy of Sciences meeting on America's Climate Choices. Kris Ebi had called me. She told me that Steve had died on an airplane from Sweden to London. She told me of her plans to go to London to retrieve Steve's body from the US Embassy and to bring him home to Terry (Root) in Palo Alto.

I came back to breakfast to break the news, but Diana Liverman had aready heard. We all left the table immediately. I don't think that anybody ate anything else that day. I was with a number of friends, and that was a comfort. - but it was up to us to deliver the news of Steve's death to colleagues from one department to the next across Washington.

When we got to the Office of Science and Technology Policy inside the White House campus first. At the beginning of an early morning that had already been scheduled, I had to tell Director John Holdren of the news. He was President Obama's Science Advisor and had been attended all of the President's 6:30 briefings.

Since I knew John, I was the designated messenger. He was stunned. But shortly after hearing the news (say, 15 seconds later), he got on the phone with the President – a direct line with no gatekeeper. "What's up?" the President asked. John told the President that Steve had died. After this short call, John reported to us that the President was also stunned.

John also told us that the President would send a sympathy note to Terry (Root). President Obama had said that there was "No need for an address – I can get that.." "Would you like me to draft something? John asked" The President said "No, I can handle that, too."

A day later, a hand-written note of sympathy was delivered to Terry at an address that none of us knew from a voice that we had all heard before. Mr. Obama had sent his sympathies with a personal touch.

Meanwhile, we slept-walked through the rest of our day. We finished our work, and I went home to the hotel - to a period of enormous grief and too much wine.

Linda got tired of that act, and sent me off to Snowmass a week later with an assignment – get over it and get well. Stop drinking. I slumped along, depressed, and I declined an invitation to attend Steve's funeral. I did not want my profound dysfunction to be the story that everyone remembered.

I was still in Snowmass when Michael Oppenheimer called. He wanted to know if I would agree to co-edit *Climatic Change* with him. He was the editor of *Climatic Change Letters*, and Springer had approached him with the challenge of putting together a team to replace Steve.

We talked on the phone in the morning dark for almost an hour while I looked out of the condo window at the Snowmass tent where I had spent many hours with Steve.

Michael and I talked some more, and finally I agreed that I would be interested. So, I said "Please keep in touch. I am sure that you have other candidates." I was trying to close the conversation. I added; "Just let me know what you decide".

Michael said something like, for reasons that I do not know and I do not want to know (except Kris Ebi and Terry and maybe even Steve were looking out for me); Michael eventually said "If you say yes, this is my first and last call." I said yes, and there you have it. CC turned out to be a lot of work, but that was not a surprise. And maybe it saved my life because there was a reason to live.

Mor than ten years later, Michael, myself, and our Deputy and Associate Deputy Editors still handle the demands of *Climate Change*. Steve would never have believed the truth. It turns out that it takes 45+ people to replace Steve, since we now handle more than 1000 new submissions and 500 revisions each year. Why? Because hundreds of people from around the world are still loyal to Steve.

Our staff does not get paid, but they frequently say that "it is the least I can do". When we ask people to become Deputy Editors (quite a bit of work), they frequently ask "hat took you so long? I would be honored."

Many times in that role, I have asked myself "what would Steve do?" Steve was a teacher, and I felt the responsibility to live up to the challenge of finding new talent from anywhere around the world. I cannot tell you how many three-page "reject before review" letters we have sent because our editorial team tries to live up to Steve's standard. And most of them end up with published papers. Many times with us, but many times elsewhere.

The last time that I saw Steve was at Jasper Ridge. As noted in chapter 13, Jasper Ridge is a Stanford research property outside of Palo Alto. It is famously noted for wildfires, poisonous snakes, cougars, and poison ivy. We were there to write uncertainty guidance for the AR5 of the IPCC, and we mostly stayed inside for good reason. Steve was bloated and having trouble standing, but he was as sharp as ever.

Terry and he had us all over for dinner on the first night of the meeting; it was a lovely, catered time – catered so that Steve and Terry could be with us and not in the kitchen. The dogs were moving from one person to another looking for droplets. Nobody talked shop. Everybody enjoyed just being in his home. We enjoyed his garden.

Steve and I spoke briefly about his upcoming speaking trip to Scandinavia and the UK in a week or so; he was looking forward to advancing iterative risk management and scientific integrity. We were all worried for him, but he would not be discouraged.

We finished our work the next morning and we all flew away home. Little did we know that that was the last time we would see him.

I got it together enough to speak at a Symposium that Terry arranged in Boulder a few years later. It was organized around the major themes of his life, and I got to talk about the history of *Climatic Change*. The major part of my talk was to report the ten most cited articles for each of the four decades of the journal's existence – based on Google Scholar and a lot of time in my hotel room because I thought of the idea on the plane.

Most of the winning authors were in the room, so an audible competition erupted as I did the David Letterman thing. "#10... #9..." and finally "#1 is..." Cheers and moans would drown me out, but I would press on.

It turns out that the Symposium was covered on the web. I found out because I heard from many authors who were not in Boulder. "My paper from 1992 had 350 citations," they would point out, "how did that not make your top ten list?" "Because #10 had 452 citations", I would answer.

This is not the story of the Symposium, though. Before I read the lists, I gave a brief history of the founding of the journal. I drew from *Climate is a Contact Sport* wherein Steve wrote about when he went to the Director of NCAR and said that he wanted to create an interdisciplinary journal about climate change. That would become *Climatic Change*. The Director said something like "If you do that, you will never receive tenure at NCAR!"

Hmmm. What do I do with that piece of news? Ignore it? No. My next slide after relating that story had three big letters – only – and a punctuation mark plus an explanation point: "WTF?!" There was silence for a second or two, but then the "crowd went wild" - at least in my memory. Really? Deny tenure to Stephen Schneider? Are you out of your minds?

Nobody from the webcast complained about the expletive, except my wife. She had been watching, and here response was... well... WTF?!!! Steve would not have objected, so I was good; and Linda got over it. I was being me for a moment just like Steve was always being Steve.

For me, Steve's influence on my life does not really show up over my annual CV. Instead, my name appears proudly with his on numbers 53, 87, and 117. All are indicative of his pathbreaking work when the IPCC was hitting its stride. This small collection of papers is also a dramatic underestimation of his influence on my thinking and writing.

As well, it is still an underestimation of the importance and pride that he and I and the entire IPCC author establishment attached to the "Responding to climate change involves a risk management approach including both adaptation and mitigation ......" finding on page 22 of #87.

He died saying those words. So will I.

#53 Ahmad, Q.K., Warrick, R., Downing, T., Nishioka, S., Parikh, K.S., Parmesan, C., Schneider, S., Toth, F., and Yohe, G., "Methods and Tools", in *Climate Change 2001: Impacts, Adaptation and Vulnerability*, Cambridge: Cambridge University Press, 2001.

The purpose of this chapter is to address several overarching methodological issues that transcend individual sectoral and regional concerns. In so doing, this chapter focuses on five related questions:

How can current effects of climate change be detected?
How can future effects of climate change be anticipated, estimated, and integrated?
How can impacts and adaptations be valued and costed?
How can uncertainties be expressed and characterized?
What frameworks are available for decision-making?

In addressing these questions, each section of Chapter 2 seeks to identify methodological developments since the Second Assessment Report (SAR) and to identify gaps and needs for further development of methods and tools.

#87 Bernstein, L., Bosch, P., Canziani, O., Chen, Z., Christ, R., Davidson, O., Hare, W., Huq, S., Karoly, D., Kattsov, V., Kundzewicz, Z., Liu, J., Lohmann, U., Manning, M., Matsuno, T., Menne, B., Metz, B., Mirza, M., Nicholls, N., Nurse, L., Pachauri, R., Palutikof, J., Parry, M., Qin, D., Ravindranath, N., Reisinger, A., Ren, J., Riahi, K., Rosenzweig, C., Rusticucci, M., Schneider, S., Sokona, Y., Solomon, S., Stott, P., Stouffer, R., Sugiyama, T., Swart, R., Tirpak, D., Vogel, C., and Yohe, G., *Climate Change 2007: Synthesis Report (for the Fourth Assessment Report of the Intergovernmental Panel on Climate Change)*, Cambridge: Cambridge University Press, 2007.

This Synthesis Report is based on the assessment carried out by the three Working Groups (WGs) of the Intergovernmental Panel on Climate Change (IPCC). It provides an integrated view of cli-mate change as the final part of the IPCC's Fourth Assessment Re-port (AR4). Topic 1 summarizes observed changes in climate and their effects on natural and human systems, regardless of their causes, while Topic 2 assesses the causes of the observed changes. Topic 3 presents projections of future climate change and related impacts un-der different scenarios. Topic 4 discusses adaptation and mitigation options over the next few decades and their interactions with sustainable development. Topic 5 assesses the relationship between adaptation and mitigation on a more conceptual basis and takes a longer-term perspective. Topic 6 summarizes the major robust findings and remaining key uncertainties in this assessment.

**#117** Smith, J.B., Schneider, S. H., Oppenheimer, M., Yohe, G., Hare, W., Mastrandrea,, M.D., Patwardhan, A., Burton, I., Corfee-Morlot, J., Magadza, C.H.D., Füssel, H-M, Pittock, A.B., Rahman, A., Suarez, A., and van Ypersele, J-P, "Dangerous Climate Change: An Update of the IPCC Reasons for Concern", *Proceedings of the National Academy of Science* 106: 4133-4137, March 17, 2009. The UNFCCC (http://unfccc.int/resource/docs/convkp/conveng.pdf) commits signatory nations to stabilizing greenhouse gas concentrations in the atmosphere at a level that "would prevent dangerous anthropogenic interference (DAI) with the climate system." In an effort to provide some insight into impacts of climate change that might be considered DAI, authors of the Third Assessment Report (TAR) of the Intergovernmental Panel on Climate Change (IPCC) identified 5"reasons for concern" (RFCs). Relationships between various impacts reflected in each RFC and increases in global mean temperature (GMT) were portrayed in what has come to be called the "burning embers diagram."

In presenting the "embers" in the TAR, IPCC authors did not assess whether any single RFC was more important than any other; nor did they conclude what level of impacts or what atmospheric concentrations of greenhouse gases would constitute DAI, a value judgment that would be policy prescriptive. Here, we describe revisions of the sensitivities of the RFCs to increases in GMT and a more thorough understanding of the concept of vulnerability that has evolved over the past 8 years. This is based on our expert judgment about new findings in the growing literature since the publication of the TAR in 2001, including literature that was assessed in the IPCC Fourth Assessment Report (AR4), as well as additional research published sinceAR4. Compared with results reported in the TAR, smaller increases in GMT are now estimated to lead to significant or substantial consequences in the framework of the 5 "reasons for concern.

# CHAPTER 21

# MENTORS

*The lede: behind every... For somebody like me to have enjoyed such a fortunate and privileged ride, let's make one thing clear – just as Gordon Ramsey on Master Chef would say – there were way too many major influences to list.*

*Here are the ones that came to mind when I was thinking about the climate problem many times in the middle of the night.*

## William Nordhaus

Bill was instrumental in bringing me into the climate wars, and keeping me there as a mentor and a colleague. Also, as will be a common theme in this chapter, he is and/or was a very good friend.

As noted in chapter 3, Bill was there at the beginning of my exploration of climate. He is the reason that I got into climate change. He and I worked together early on an Academy report, and then on the first version of DICE, and beyond.

When he won his named Prize, he said to me, "I caught up"; and then he invited me to be a guest at the ceremonies in Stockholm (see Chapter 19).

Being awarded a small part of the Peace Prize was one thing even if I lost the lottery to go to Oslo. But his was named and the ceremony was in Stockholm. Bill showed me *the top of the world!* No doubt about it. I will forever be grateful.

Bill has always seemed to be proud of me, and that makes me smile even when I cannot sleep at night. Without his influence, my life would have been entirely different and I would have wasted many years in obscurity writing esoteric papers on micro-scale decision-making under uncertainty. Had I done that, there would have been no answers to "so what do you do?" at a Christmas party, or later when my granddaughters ask "what did you do to save our planet?" When you follow Bill, the answer is "I was there, we are still there, and we continue to protect humanity from climate change!"

The story of the National Academy report on Changing Climate in 1982 has already been told above in Chapter 3. Bill and I had worked to create and to explain probabilistic scenarios of carbon emissions and atmospheric concentrations. William Nurenburg was the chair of the Committee, and he was pleased. Also on the Committee were several future Nobel Laureates. We did not know that at the time, and hardly anybody makes mention. but the Report is still important because it is still cited.

Based on that work, Bill arranged for me to attend a meeting of the International Energy Workshop (IEW) at IIASA in Laxenburg, Austria. It was my first trip abroad at 27 years of age. Really – my first one. Linda came. We stayed in Vienna in a comfortable, small hotel next to the OPEC secretariat offices. On a day early in our stay before the meetings began, Bill took the time to take us both to Demel in the Inner Stadt.

He walked around Demel pointing to this pastry and that to show us how beautiful they were. We were very impressed. We admired every one. We did not understand that we were buying every one. A very attentive attendant thought that Bill adding to a big order and presented Bill with a large tray of real pastries for his approval when we walked to the front door to leave. Bill, myself, and Linda all declined – big misunderstanding we thought! We all walked out of the store and down the sidewalk, but we were followed by much commotion. I took a while to sort things out.

I have been to Vienna many times since then, but it took me a decade before I could go back into Demel. I was sure that they would remember me. I do not know if either Bill or Barbara ever returned for another tryl

So, let me finish by asking the one question that I had been asking myself for the 20 years leading up to 2018: "When would the Nobel Prize in Economics show up on his doorstep?"

Who asks such a question? Nobel Prizes don't just happen, and your cannot feel diminished if it never happens. But in my head, it always seemed perfectly natural for William Nordhaus. Finally, we found out that the answer to my question was 2018. He was then, and so deserving for so many reasons.

# Kristie Ebi

Kris is an honest and sincere soul who is generous to a fault. She is very, very smart (and street smart – they are not the same thing). She has been an anchor for my vulnerable sanity and productivity over decades through IPCC and beyond (see https://globalhealth.washington.edu/faculty/kristie-ebi). A source of stability and love for me and my family, she travels like Steve did. I worry.

Kris and I have shared experiences at the National Academy of Sciences (including the Institute of Medicine), IPCC, National Climate Assessment, and Climatic Change. We have been meeting each other at various place around the world for more than three decades and we have many co-authored papers. Occasionally, we actually meet close to where one or the other of us lives. She has seen my highs and my lows, and was always been there for me. I have shared her highs and lows, too; and I hope that I have always been there for her – though I am apologetically not sure.

She travels too much. Always has. Always will. Anybody who needs her help to cope with climate risk will always find her willing to come asap. She replaced Tom Malone for the game – "where in the world is (Kris Ebi when the kids were growing up)...?" We had two global maps to keep things straight.

These days it is "is Kris home, or is she traveling?"

Happily, now that she has a grandson in Seattle, she is home more often than not - but not always. She and I share "grands" stories; and before then, various stories about Caroline, my Bishon Friese.

It turns out that 10-pound Caroline had her own hawk. The hawk fed her one winter because she looked hungry in her snowed-in place of "business"; he dropped a dead squirrel for her to take inside. Why? I remember the day that she helped the hawk catch a squirrel earlier that year. Caroline was allergic to squirrels. She was doing her business, saw the squirrel and erupted with a loud bark. One ill-trained squirrel heard her bark and stood up to see the source of the commotion. The hawk swept over Caroline (and me) and picked up the now very vulnerable squirrel. It was the beginning of an unexpected friendship.

Kris's favorite line when my friends and I get out of our sandboxes is "have you talked to an epidemiologist?" Usual not, though she knows that we are just looking for parallels and synergies. We have actually coauthored on these.

Let me tell you one story to supplement the many papers that populate my yearly CV. One of her daughters, Katie, worked with my daughter Courtney for Susan Sweeney for many years in Snowmass to make the meetings work smoothly. One night at dinner, under the tent, Kris and I were chatting (probably about a presentation or an upcoming collaboration, but certainly with some wine) and Katie came up to talk to her mom. "Yo, Mama!", she said. "That be us," I replied. I could not resist. Everyone close to us laughed out loud, and Katie became my friend for life because I was so cool (???).

Kris always pushed health issues in impacts meetings. For a very long time, health was the last impact to be considered when everyone was collecting baggage to catch a flight. I like to think that our work to map the "determinants of adaptive capacity" to the "precursors of public health" helped bring health closer to the fore if peoples' consciousness (please see #64). It has been a long climb up a steep hill over many years, but our other collaborations include numbers 72, 81, 151, 170, and 175

Perhaps we were half way up that hill when the Trump Administration declared climate impacts were all a hoax.

With their persistent and generous funding of the NIH, perhaps Congress saw something different. Perhaps the real overlap is that the climate change community finally saw its problem in terms of risk management, adaptation and mitigation. It turns out that public health had been viewing its mandate through the risk management lens for nearly a century.

An issue, though, is that medicine has created flow charts to organize standard decisions at anticipated outcome bifurcations designed to handle the "normal" (in a statistical sense), patient. As Steve Schneider emphasized from personal and successful experience, no patient is normal. My daughter Mari, the pediatric oncologist Lasker fellow at the NIH agrees. Turns out that she learned some statistics along the way to that lofty position.

**#64** Yohe, G. and Ebi, K. "Approaching Adaptation: Parallels and Contrasts between the Climate and Health Communities" in *Integration of Public Health with Adaptation to Climate Change: Lessons Learned and New Directions* (Ebi, K., Smith, J. and Burton, I., eds.), The Netherlands: Taylor and Francis, 2005.

**#81** Tol, R., Ebi, K., and Yohe, G., "Infectious Disease, Development and Climate Change: A Scenario Analysis", *Environment and Development Economics* 12: 687-706, 2007.

This study the effects of development and climate change on infectious diseases in Sub-Saharan Africa. Infant mortality and infectious disease are closely related, but there are better data for the former. In an international cross-section, per capita income, literacy, and absolute poverty significantly affect infant mortality. We use scenarios of these three determinants and of climate change to project the future incidence of malaria, assuming it to change proportionally to infant mortality. Malaria deaths will first increase, because of population growth and climate change, but then fall, because of development. This pattern is robust to the choice of scenario, parameters, and starting conditions; and it holds for diarrhea, schistosomiasis, and dengue fever as well. However, the timing and level of the mortality peak is very sensitive to assumptions. Climate change is important in the medium term, but dominated in the long term by development. As climate can only be changed with a substantial delay, development is the preferred strategy to reduce infectious diseases even if they are exacerbated by climate change. Development can, in particular, support the needed strengthening of disease control programs in the short run and thereby increase the capacity to cope with projected increases in infectious diseases over the medium to long term. This conclusion must, however, be viewed with caution.

## Thomas Malone (1917-2013)

Tom was the catalyst who led me to broaden my international engagement. https://www.nasonline.org/publications/biographicalmemoirs/memoir-pdfs/malone_thomas_pdf tells you how that happened.

He was my point of entry for much of my engagement with the global world, but he was also the anchor for my children's engagement to the rest of the world beyond Portland, CT.

The phone would ring out our house. One or the other of them would answer. Whichever would hear: "Hello. This is Tom Malone calling from the (fill in the blank – e.g., Singapore – airport). This was way before caller ID. Is your father there?" One daughter would get me to the phone and the other both would to the map of the world to play the "Where in the world is Tom Malone" game. "Quick. Find him on the map," was the objective.

"OK! I found Singapore. Wow! He is far away!".

Tom was usually far away. In fact, he racked up so many travel miles on his credit card that he and Rosalie could, with paying a penny, plan and enjoy together a 40 day trip around the world.

Tom never told me that he was a MacArthur Fellow. Turns out I have known six Fellows, including a recent Wesleyan alum, but Tom was probably the first. Or maybe Bill Clark from his review of my chapter with Bill Nordhaus in *Changing Climate*.

Our first significant excursion abroad together was to attend the Second World Climate Conference (SWCC) in Geneva (October 19-November 7 in 1990). He was chair of Working Group 12 (there were multiple break out groups), and I volunteered to his staff. That was a good idea. I took notes every day, showed them at breakfast the next day, and ultimately wrote the penultimate version of our report. He edited what became our joint proposal to create an integrated collection of research and training institutes scattered around the world... D.C., Bangladesh, etc.

Our report was accepted by the SWCC. I was thrilled and hooked on this level of engagement with an international community. A few months later, a session that Tom hosted in Bellagio came up with a formal proposal for our idea in Geneva. They created an institution that became START – SysTem of regional networks for Analysis, Research and Training. The program still exists, working to expand capacity in Africa and Asia. It has changed small and large parts of the world and the lives of tens of thousands of people each year.

Toward the end of his life, I was successful in nominating Tom for an honorary degree from Wesleyan. I missed my cousin Angela's wedding (sorry – can you forgive me?) because I had the honor of introducing him at Commencement and placing OUR WESLEYAN hood on his shoulders. Tears all around, there were.

Tom and I had collaborated on several papers (#23, #26, and #58); but the collaboration went far beyond the paper or that. His concern about the planet was infectious. I learned from him that you can never do too little, even if you have to travel too much. But sometimes you accomplish a lot!

**#59** Malone, T., and Yohe, G., "Knowledge Partnerships for a Sustainable, Equitable and Stable Civilization", *Journal of Knowledge Management*, 6: 368-378, 2002.
Continued exponential and asymmetric growth in both population and individual economic productivity would propel world society along a path that is environmentally unsustainable, economically inequitable, and hence socially unsustainable. Revolutionary developments in communications technologies can, however, enable partnerships among scholarly disciplines and among societal institutions to harness rapidly expanding human knowledge to pursue goals in both population and individual productivity that would least to a sustainable, equitable, and stable world society. Several scenarios are presented to illustrate the promise of cooperative efforts to pursue this vision and to highlight some obstacles that persist.

# Thomas Wilbanks (1938-2017)

Tom Wilbanks was a true southern gentleman with smarts, backbone, and grace (please see https://www.ccsi.ornl.gov/sites/default/files/wilbanks_bio.pdf) . Tom and I followed each other around the world. That is to say that we shared a hotel breakfast on six continents; I was never sure who got their first, but an early breakfast after a night of work that ended around 2PM local time was a common bind.

Especially, back and forth to Washington from CT and TN for three decades, we worked at the same places as Kris Ebi: IPCC meetings, National Academy of Sciences meetings, National Climate Assessment meetings, and the like.

He always traveled more than I did, but that was through my choice, balancing family and obligations. I declined more frequently than most of my colleagues. Almost declined the White House, as you know if you have read this far linearly.

When we did end up at the same meeting (and the same hotel which happened frequently, we would share thoughts about the meetings – what was about to come and what had just happened. He would give me advice about how to respond. Invaluable and productive would that advice be.

We would, though, share life experiences. He about hi "back door clearance" at the Pentagon; me about Mari and Courtney (daughters who went to Geno Auriemma's basketball camps.

Tom was an avid Lady Vol fan, and I was not. I won quite a few dollar bets on games when Tennessee and UCONN used to play three times a year.

I think that we learned a lot from each other; I certainly learned a lot from him. But what I remember most is his humanity.

In one episode, Tom was retiring from chairing a standing National Academy Committee on climate adaptation and resilience, or something like that. He had chaired his last meeting, and we arranged to have dinner. I had arranged to have a pewter Jefferson Cup from Williamsburg engraved to commemorate his service and his retirement, and I gave it to him at dinner. No fanfare or hype. I just wanted him to know how much he meant to me.

One vivid memory - Linda and I had a quiet dinner with Tom just outside of Merida, Mexico (we were there for an IPCC meeting). It was one of the highlights of what turned out to be a wonderful (top 5 in my life) trip. There was a beautiful view. Dinner was relaxed and wonderful. We talked about family, basketball, and NOT work. A few hours later, we all took a carriage ride back to the hotel campus. We passed Ian Burton and Barry Smit having a drink on a sidewalk table. Linda shouted "Hola!" Ian shouted back "Olay!" We waved and all smiled.

We all laughed the next morning at a shared table for breakfast. This was the IPCC community at its best – save the planet one chapter at a time, but make life-long friends as you go.

**#88** Janetos, A., Balstad, R., Apt, J., Ardanuy, P, Friedl, R., Goodchild, M., Macauley, M., McBean, G., Skole, D., Welling, L., Wilbanks, T., and Yohe, G., "Earth Science Applications to Societal Benefits", in *Earth Science and Applications from Space: National Imperatives for the Next Decade and Beyond,* National Research Council, Washington, D.C.: The National Academies Press, 2007.
The theme of this chapter is the urgency of developing useful applications and enhancing benefits to society from the nation's investment in Earth science research. Accomplishing this objective requires an understanding of the entire research-to-applications chain, which includes generating scientific observations, conducting research, transforming the results into useful information, and distributing the information in a form that meets the requirements of both public and private sector managers, decision makers, policy makers, and the public at large.

## Jerry Melillo

Jerry came into my life strongly but quite late, even though I had known of him for many decades (see https://www.nasonline.org/member-directory/members/3435.html). Our close professional association was born of my participation as one of his two vice-chairs for the Third National Climate Assessment (NCA3) for President Obama (T.C. Richmond was the other). He gently taught us about how to behave, how to lead by bringing others to their own conclusions [across the 44-member National Climate Assessment Development and Advisory Committee (NCADAC)], and how to make the world a better place.

What did I contribute. I think that I helped him understand how to work a room to bring them to consensus according to the rules described in Chapter 8, and why that would be a good thing.

A giant in the scientific world for his research, Jerry had vast experience in leading and organizing large and small groups of "strong, scientific and private industry cats". Herding them was always a challenge, but he taught me that listening and responding respectfully was the key. All I did was bring the definition of "consensus" into the decision-making process.

NCA3 decided early on, with my encouragement, that the NCADAC would make its decisions and assert its findings on the basis of consensus. I had IPCC experience in managing that, so he let me lead meetings where we would seek consensus. To be clear, that meant that for a particular word or sentence or conclusion or instruction... anyone in the room could object if he or she could propose an alternative.

It took a while for Committee members to understand their responsibilities in this process. Early decisions on even a single well selected word in plenary took hours. Eventually, finding consensus on more important words worked efficiently. In the end, the final NCA3 report and their two derivative documents passed with very little drama.

That is, except for one source of drama described in our history, The NCA3 had divided the US into eight, sometimes very large, regions. Then there was that source of drama. As reported above, the Administration, in an effort lead by John Podesta, wanted to release state by state "two-pagers" derived from the report. Jerry and T.C. and I, agreed with our small Secretariat that we could not do that.

NCA3 documentation did not have sufficient skill to do that credibly. Jerry, T.C., and I decided that (to protect the entire report from being shot down with false statements in the ancillary state by state releases), we would withhold our names from the cover pages of the entire report if the White House prevailed in this project.

Long story short, as you know, the White House blinked, and the NCA3 was released at 8:30 the next morning – on time and by consensus. Jerry went to the Rose Garden with the President, and the rest of us watched the electronic release of our 1600-page report (#162).

By 9:00, our site had accommodated 20,000 hits with ease. In the press briefing that afternoon in the Eisenhower Executive Office Building auditorium, John Podesta declared something like: "see, this White House can actually roll out something electronic and handle all comers". Sadly, the less than perfect experience with the Affordable Care Act was still in the public consciousness. Both were similarly important

**#160** Melillo, J, Richmond, T.C., and Yohe, G., and many others, *Third National Climate Assessment,* the White House, United States Global Change Research Program and the National Oceanic and Atmospheric Administration, May 6, 2014 (Overall report, *Highlights, Overview,* and Regional Reports available at www.nca2014.globalchange.gov).

# Alan Manne (1925-2005)

Alan was a kind demanding mentor.

When we met in Snowmass, Alan was already a leading scholar in his field. (see https://www.msande.stanford.edu/ people/alan-manne). We were in Snowmass. He was then expanding his influence to integrated assessment (of climate change) through the creation and evolution of a model named MERGE. Through this model, I met and came to enjoy the friendship of Richard Richels – one of Alan's students – and many others.

Alan was a gentle tutor when I was new to the game. Alan became a dogged, constructive skeptic of my work when he had promoted me in his mind to a more mature level where the expectations were higher. He did not suffer educated fools well, and so achieving that threshold made me vulnerable to his questioning, even in public.

That was fine, because I learned quickly that I should always be prepared when I was working on a presentation particularly for Snowmass where Alan would be in the audience. Just anticipating what he would ask made my work and my presentations better, and so it made my contribution to science better.

Sometimes I would put my anticipation in the presentation, but other times I would just be ready for the question that I knew would be coming. By the end, I could get through a talk without a question or, if I wanted to, look smart when I answered Alan's queries.

This sounds like manipulating the audience, but he knew what I was doing. You cannot imagine how much I learned about my own research by preparing for Alan.

Bill Nordhaus was in the same category, but our conversations were always more private (usually over lunch at Clark's Dairy in New Haven).

Just like contemplating what Waino Fillbacks, a church elder, made me do when I was President of the First Congregational Church of Portland. "What will he ask?" was always the source of the last round of revisions of any presentation and my expectations about first round of questions at the beginning of the next thought exercise.

Alan and I worked together to frame modeling exercises for the Energy Modeling Forum (EMF) uncertainty group. The exercises were designed to support statistically based model comparisons. We discovered one important tendency. Given the opportunity of pick driving variables instead of using specified characterizations, the distribution of outcomes like carbon emissions were smaller when everyone used the same inputs.

Modeler's choice, therefore, displayed modelers' instincts not to stand out from the crowd. And so, distributions of outcomes from modelers' choices were NOT accurate measures of uncertainty.

This insight was a big deal.

My memories of Alan are not confined to academic issues. He arranged for my daughter (Courtney) and me to go horseback riding at Maroon Bells. I was on my own without instruction, but Courtney got personal attention. Within 30 minutes, she was comfortable on a horse that would carry her along some rugged countryside for a few hours.

Alan and I, in on of our last times together, enjoyed croque-monsieur in a Brussels café (after an EMF meeting). We ate and chatted about important people in our lives before we slowly wandered back to the hotel. We picked up presents for the people at home along the way. He bought chocolate. I bought lace. It took a while, but we enjoyed the smiles from and conversations with the locals.

My most precious memory of my time with Alan was taking him as the guest of our family to a concert in the Musikverein in Vienna. He fell asleep for a bit, but he awoke in time to buy us all champagne at intermission. Mari and Courtney were in heaven.

Then there was the meeting at a Subway (restaurant) in Vienna – we met there by accident in the dark, way after dinner time, because we were both looking for a touch of home – an Italian grinder for me and tuna salad for him. Needless to say, I miss him. His number is also still on my cell phone list.

The list of derivative papers from my collaboration with him is long: starting at the beginning of my career, numbers 12, 15, 19, 21, 25, 31, 32, 34, 42, and 68, to name a few. Nearly everything that I wrote was presented before him in those days, and they were all were better because of his anticipated and actual scrutiny.

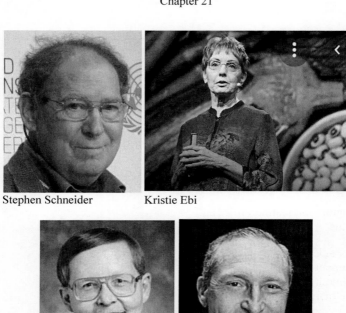

Stephen Schneider          Kristie Ebi

Thomas Wilbanks          Alan Manne

Jerry Millilo              Thomas Malone

**Figure 20-1 Selected mentors over a lifetime**

# APPENDIX A

# MY JANUARY 2023 CV WITH EXPANDED NUMBERED REFERENCES FOR THE TEXT

**Gary Wynn Yohe**                              **January 2023**

Huffington Foundation Professor of Economics and Environmental Studies, Emeritus

Wesleyan University
Department of Economics
238 Church St.
Middletown, CT 06459 USA

E-mail: gyohe@wesleyan.edu
Research outreach: https://researchoutreach.org/articles/climate-change-40-years-career/

**Education:**

1970 B.A. University of Pennsylvania, Mathematics.
1971 M.A. State University of New York at Stony Brook, Mathematics.
1974 M.Phil. Yale University, Economics.
1975 Ph.D. Yale University, Economics.

**Fellowships and Awards**

Phi Beta Kappa, University of Pennsylvania.

Yale University Fellowship.

Elected member of Connecticut Academy of Science and Engineering.

Elected member-at-large of Sigma Xi.

Co-recipient of the 2007 Nobel Peace Prize as a senior member of the Intergovernmental Panel on Climate Change for the Third and Fourth (and subsequently the Fifth) Assessment Reports.

Honorary Chair of Connecticut United Nations Day by decree of Governor Malloy.

Wesleyan's First Faculty Research Prize, September 4, 2018.

Invited guest of William Nordhaus at his 2018 Nobel Prize Ceremony, Stockholm, December 6-12, 2018.

**Selected Publications:** A * before a name in the author list identifies a Wesleyan undergraduate student who was full-fledged co-author of the listed paper. The article names are hyperlinked to an online available version at gyohe.faculty.wesleyan.edu. If there is no hyperlink to the article, it is not available - this document will provide a source where the article can be purchased. One, two, or three checks ($\sqrt{\ }$) indicates more than 250, 500, or 1000 citations, respectively; four checks indicates more than 10,000 citations for #60, Parmesan and Yohe (2003).

See also https://www.researchgate.net/profile/Gary-Yohe .

## 1976

1.      **Yohe, G.**, "Substitution and the Control of Pollution--A Comparison of Effluent Charges and Quantity Standards Under Uncertainty," *Journal of Environmental Economics and Management* **3**: 312-324, 1976.
2.      **Yohe, G.**, "Polluters' Profits and Political Responses: Direct Control versus Taxes", *American Economic Review* **66**: 981-982., 1976.

## 1977

3.      **Yohe, G.**, "Single-Valued Control of a Cartel Under Uncertainty--A Multifirm Comparison of Prices and Quantities" *Bell Journal of Economics* **8**: 97-111, 1977.
4.      **Yohe, G.**, "Single-valued Control of an Intermediate Good Under Uncertainty--A Comparison of Prices and Quantities" *International Economic Review* **18**: 117-133, 1977.
5.      Ogura, S. and **Yohe, G.**, "The Complementarity of Public and Private Capital and the Optimal Rate of Return to Government Investment," *Quarterly Journal of Economics* **91**: 651-662, 1977.

**1978**

6.     **Yohe, G.**, "Towards a General Comparison of Price Controls and Quantity Controls Under Uncertainty" *Review of Economic Studies* **45**: 229-238, 1978.

**1979**

7.     **Yohe, G.**, *A Comparison of Price Controls and Quantity Controls Under Uncertainty*, New York: Garland Publishing Co. Inc., 1979. (Published in a series of 24 Outstanding Dissertations in Economics).

8.     Karp, G. and **Yohe, G.** "The Optimal Linear Alternative to Price and Quantity Controls in the Multifirm Case" *Journal of Comparative Economics* **3**: 56-65, 1979.

**1980**

9.     **Yohe, G.**, "Current Publication Lags in Economics Journals," *Journal of Economic Literature* **18**: 1050-1055, 1980.

**1983**

10.    √ Nordhaus, W. and **Yohe, G.**, "Future Carbon Dioxide Emissions from Fossil Fuels," in *Changing Climate: Report of the Carbon Dioxide Assessment Committee*, Washington: National Research Council, 1983.

11.    Exercises and Applications for Microeconomic Analysis (1st Edition) New York: Norton and Company, 1983 (2nd and 3rd editions in 1986 and 1993).

**1984**

12.    **Yohe, G.**, "The Effects of Changes in Expected Near-term Fossil Fuel Prices on Long-term Energy and Carbon Dioxide Projections" *Resources and Energy* **6**: 1-20, 1984.

13.    **Yohe, G.**, "Constant elasticity of substitution production functions with three or more inputs: and approximation procedure", *Economics letters, 1984.*

**1985**

14.     **Yohe, G.**, Study Guide to Accompany Samuelson/Nordhaus 'Economics', 12th edition, New York: McGraw Hill Publishers, 1985 (13th and 14th editions in 1989 and 1992).

**1986**

15.     **Yohe, G.**, "Evaluating the efficiency of long-term forecasts with limited information: Revisions in the IEW poll responses", *Resources and Energy* **8**:331-339, 1986.

**1987**

16.     **Yohe, G.** and MacAvoy, P., "The Practical Advantages of Tax Over Regulatory Policies in the Control of Industrial Pollution", *Economics Letters* **25**: 177-182, 1987.

**1989**

17.     **Yohe, G.**, "More on the Properties of a Tax Cum Subsidy Pollution Control Strategy," *Economic Letters* **31**: 193-198, 1989.
18.     **Yohe, G.**, "The Cost of Not Holding Back the Sea – Economic Vulnerability", *Ocean Shoreline Management* **15**: 233-255, 1989.

**1991**

19.     **Yohe, G.**, "Uncertainty, Global Climate and the Economic Value of Information", *Policy Sciences* **24**: 245-269, 1991.
20.     **Yohe, G.**, "The Cost of Not Holding Back the Sea – Towards a National Sample of Economic Vulnerability", *Coastal Management* **18**: 403-431, 1991.
        https://www.tandfonline.com/doi/abs/10.1080/08920759009362123
21.     **Yohe, G.**, "Selecting 'Interesting' Scenarios with which to Analyze Policy Response to Potential Climate Change" *Climate Research* **1**: 169-177, 1991.
22.     √ √ Titus, J.G., Park, R.A., Leatherman, S.P., Weggel, J.R., Greene, M.S., Mausel, P.W., Brown, s., Gaunt, G., Threhan, M. and **Yohe, G.**, "Greenhouse Effect and Sea Level Rise: the Cost of Holding Back the Sea", *Coastal Management* **19**: 171-204, 1991.

**1992**

23.   **Yohe, G.** and Malone, T. "Toward a General Methodology with Which to Evaluate the Social and Economic Impact of Global Change," *Global Environmental Change* **3**: 101-110, 1992.

24.   **Yohe, G.**, "Carbon Emissions Taxes: Their Comparative Advantage under Uncertainty," *Annual Review of Energy* **17**: 301-326, 1992.

25.   **Yohe, G.**, "Imbedding Dynamic Responses with Imperfect Information into Static Portraits of the Regional Impact of Climate Change" in *Economic Issues in Global Climate Change* (Reilly, J. and Anderson, M., eds.), Oxford: Westview Press, 1992.

**1993**

26.   **Yohe, G.**, "Equity and Efficiency in the Clinton Energy Tax: Some Early Thoughts from First Principles", *Energy Policy* **21**: 953-957, 1993.

27.   **Yohe, G.**, "EMF 12.2: Sorting Out Facts and Uncertainties in Economic Response", *Energy Modeling Forum*, Stanford University, January 1, 1993.

**1995**

28.   **Yohe, G.** and *Garvey, B., "Incorporating Uncertainty and Nonlinearity into the Calculus of an Efficient Response to the Threat of Global Warming", *International Journal of Global Energy Issues* **7**: 34-47, 1995.

29.   **Yohe, G.**, Neumann, J. and Ameden, H., "Assessing the Economic Cost of Greenhouse Induced Sea Level Rise: Methods and Applications in Support of a National Survey", *Journal of Environmental Economics and Management* **29**: S-78-S-97, 1995.

**1996**

30.   √ **Yohe, G.**, Neumann, J., Marshall, P., and Ameden, H., "The Economic Cost of Greenhouse Induced Sea Level Rise in the United States", *Climatic Change* **32**: 387-410, 1996.

31.   **Yohe, G.**, and *Wallace, R. "Near Term Mitigation Policy for Global Change Under Uncertainty - Minimizing the Expected Cost of Meeting Unknown Concentration Thresholds", *Energy Modeling Assessment* **1**: 47-57, 1996.

32.    **Yohe, G.**, "Exercises in Hedging Against Extreme Consequences of Global Change and the Expected Value of Information" *Global Environmental Change* **6**: 87-101, 1996.

**1997**

33.    **Yohe, G.** and Neumann, J., "Planning for Sea-level Rise and Shore Protection under Climate Uncertainty", *Climatic Change* **37**: 243-70, 1997.

34.    **Yohe, G.**, "Uncertainty, Short Term Hedging and the Tolerable Windows Approach" *Global Environmental Change* **7**: 303-315, 1997.

35.    Schellnhuber, J. and **Yohe, G.**, *Comprehending the Economic and Social Dimensions of Climate Change by Integrated Assessment*, World Climate Research Program, Geneva, 1997.

**1998**

36.    **Yohe, G.**, and Cantor, R., "Economic Activity" in *Human Choice and Climate Change*, Volume 3 (Chapter 1), Washington: Battelle Press, 1998.

37.    √ **Yohe, G.** and Schlesinger, M., "Sea Level Change: The Expected Economic Cost of Protection or Abandonment in the United States", *Climatic Change* **38**: 447-472, 1998.

38.    **Yohe, G.**, Malinowski, T., Yohe, M., "Fixing Global Emissions: Choosing the Best Target Year", *Energy Policy* **26**: 219-231, 1998.

**1999**

39.    **Yohe, G.**, Neumann, J. and Marshall, P., "The Economic Damage Induced by Sea Level Rise in the United States" in *The Impact of Climate Change on the United States Economy* (Mendelsohn, R. and Neumann, J., eds.), Cambridge: Cambridge University Press, 1999.

40.    **Yohe, G.** and Schimmelpfennig, D., "Vulnerability of Agricultural Crops to Climate Change: A Practical Method of Indexing" in *Global Environmental Change and Agriculture: Assessing the Impacts*, Edward Elgar Publishing, 1999.

41.    **Yohe, G.**, *Jacobsen, M., and *Gapotochenko, T., "Spanning 'Not Implausible' Futures to Assess Relative Vulnerability to Climate Change and Climate Variability", *Global Environmental Change* **9**: 233-249, 1999.

42.    **Yohe, G.**, "The Tolerable Window Approach - Lessons and Limitations", *Climatic Change* **41**: 283-295, 1999.

43.    **Yohe, G.** and Dowlatabadi, H., "Risk and Uncertainties, Analysis and Evaluation: Lessons for Adaptation and Integration", *Mitigation and Adaptation Strategies for Global Change* **4**: 319-329, 1999.

**2000**

44.    **Yohe, G.**, "Assessing the Role of Adaptation in Evaluating Vulnerability to Climate Change", *Climatic Change* **46**: 371-390, 2000.

45.    **Yohe, G.**, and Toth, F., "Adaptation and the Guardrail Approach to Tolerable Climate Change", *Climatic Change* **45**: 103-128, 2000.

46.    *Microeconomics* (10th & 11th editions, with Edwin Mansfield), Norton and Company, 2000 & 2003.

47.    Neumann, J., **Yohe, G.**, Nicholls, R., Manion, M., "Sea-Level Rise & Global Climate Change: A Review of Impacts to U.S. Coasts", Pew Center on Global Climate Change, Washington, DC, 2000.

48.    **Yohe, G.**, "Integrated Assessment of Climate Change – the Next Generation of Questions", *Climate Impact Research: Why, How, and When?"* Akademie Verlag GmbH, Berlin: Joint International Symposium, 2000; see 1997.

49.    **Yohe, G.**, Montgomery, D., and Balistreric, E., "Equity and the Kyoto Protocol: Measuring the Distributional Effects of Alternative Emissions Trading Schemes", *Global Environmental Change* **10**: 121-132, 2000.

**2001**

50.    √ **Yohe, G.**, "Mitigative Capacity – The Mirror Image of Adaptive Capacity on the Emissions Side", *Climatic Change* **49**: 247-262, 2001.

51.    Wassserman, H.* and **Yohe, G.**, "Segregation and the Provision of Spatially Defined Local Public Goods", *American Economist* **45**: 13-24, 2001.

52.    Strzepek, K., Yates, D., **Yohe, G.**, Tol, R. and *Mader, N., "Constructing "Not Implausible" Climate and Economic Scenarios for Egypt", *Integrated Assessment* **2**: 139-157, 2001.

53.    Ahmad, Q.K., Warrick, R., Downing, T., Nishioka, S., Parikh, K.S., Parmesan, C., Schneider, S., Toth, F., and **Yohe, G.**, "Methods and

Tools", in *Climate Change 2001: Impacts, Adaptation and Vulnerability,* Cambridge: Cambridge University Press, 2001.

54. Smit, B., Pilifosova, O., Burton, I., Challenger, B., Huq, S., Klein, R.J.T., and **Yohe, G.,** "Adaptation to Climate Change in the Context of Sustainable Development and Equity", in *Climate Change 2001: Impacts, Adaptation and Vulnerability,* Cambridge: Cambridge University Press, 2001.

55. Smith, J., Schellnhuber, J., Mirza, M., Fankhauser, S., Leemans, R., Erda, L., Ogallo, L., Pittock, B., Richels, R., Rosenzweig, C., Safriel, U., Tol, R.S.J., Weyant, J., **Yohe, G.,** "Vulnerability to Climate Change and Reasons for Concern: A Synthesis", in *Climate Change 2001: Impacts, Adaptation and Vulnerability,* Cambridge: Cambridge University Press, 2001.

56. Banuri, T., Weyant, J., Akumu, G., Najam, A., Rosa, L., Rayner, S., Sachs, W., Sharma, R., **Yohe, G.,** "Setting the Stage: Climate Change and Sustainable Development", in *Climate Change 2001: Mitigation,* Cambridge: Cambridge University Press, 2001.

**2002**

57. √ √√ **Yohe, G.** and Tol, R., "Indicators for Social and Economic Coping Capacity – Moving Toward a Working Definition of Adaptive Capacity", *Global Environmental Change* **12**: 25-40, 2002.

58. **Yohe, G.** and Schlesinger, M., "The Economic Geography of the Impacts of Climate Change", *Journal of Economic Geography* **2**: 311-341, 2002.

59. Malone, T., and **Yohe, G.,** "Knowledge Partnerships for a Sustainable, Equitable and Stable Civilization", *Journal of Knowledge Management,***6**: 368-378, 2002.

**2003**

60. √ √ √ √ Parmesan, C. and **Yohe, G.,** "A Globally Coherent Fingerprint of Climate Change Impacts across Natural Systems", *Nature* **421**, 37-42, January 2, 2003.

**2004**

61. √ **Yohe, G.,** Andronova, N. and Schlesinger, M., "To Hedge or Not Against an Uncertain Climate Future", *Science,* **306**: 415-417, October 15, 2004.

62. Manning, M.R., Petit, M., Easterling, D., Murphy, J., Patwardhan, A., Rogner, H-H, Swart, R. and **Yohe, G.** (eds.), IPCC Workshop on Describing Scientific Uncertainties in Climate Change to Support

Analysis of Risk and of Options: Workshop report. Intergovernmental Panel on Climate Change (IPCC), Geneva, 2004.

63.    **Yohe, G.**, "Some Thoughts on Perspective", *Global Environmental Change* **14**: 283-286, 2004.

**2005**

64.    **Yohe, G.** and Ebi, K. "Approaching Adaptation: Parallels and Contrasts between the Climate and Health Communities" in *Integration of Public Health with Adaptation to Climate Change: Lessons Learned and New Directions* (Ebi, K., Smith, J. and Burton, I., eds), Taylor and Francis, The Netherlands, 2005.

65.    **Yohe, G.**, "Climate Policy in an Adapting World", *Geotimes* **50**: 26-29, 2005.

66.    **Yohe, G.**, Schlesinger, M. and Andronova, N., "Reducing the Risk of a Collapse of the Atlantic Thermohaline Circulation" *Integrated Assessment* **20**: 1-17, 2005, ISSN: 1389-5176.

**2006**

67.    **Yohe, G.**, "Representing Dynamic Uncertainty in Climate Policy Deliberations", *Ambio* **35**: 89-91, 2006.

68.    **Yohe, G.**, Malone, E., Brenkert, A., Schlesinger, M., Meij, H. and Xing, X., "Global Distributions of Vulnerability to Climate Change", *Integrated Assessment Journal* **6**: 35-44, 2006.

69.    **Yohe, G.**, Elizabeth Malone, E., Antoinette Brenkert, A., Schlesinger, M., Meij, H., Xing, X., and Lee, M., "A Synthetic Assessment of the Global Distribution of Vulnerability to Climate Change from the IPCC Perspective that Reflects Exposure and Adaptive Capacity", 2006 CIESIN, Columbia University, New York, http://sedac.ciesin.columbia.edu/mva/ccv/.

70.    Schlesinger, M., **Yohe, G.**, Yin, J., Andronova, N., Malyshev, and Li, B., "Assessing the Risk of a Collapse of the Atlantic Thermohaline Circulation" in *Avoiding Dangerous Climate Change* (Schellnhuber, H.J., Cramer, W. Nakicenovic, N. Wigley, T., and Yohe, G. eds.), Cambridge: Cambridge University Press, 2006.

71.    Tol, R. and **Yohe, G.**, "On Dangerous Climate Change and Dangerous Emission Reduction" in *Avoiding Dangerous Climate Change* (Schellnhuber, H.J., Cramer, W. Nakicenovic, N. Wigley, T., and Yohe, G. eds.), Cambridge: Cambridge University Press, 2006.

72.   **Yohe, G.**, Adger, N., Dowlatabadi, H., Ebi, K., Huq, S., Moran, D., Rothman, D., Strzepek, K., and Ziervogel, G., "Recognizing Uncertainty in Evaluating Responses", in *Ecosystems and Human Wellbeing: Volume 3 – Policy Responses,* New York: Island Press, 2006.

73.   **Yohe, G.**, "Some Thoughts on the Estimates Presented in the *Stern Review* – An Editorial*", Integrated Assessment Journal* **6**: 66-72, 2006, ISSN: 1389-5176.

74.   √ Tol, R. and **Yohe, G.**, "A Review of the *Stern Review*", *World Economy* **7**: 233-250, 2006.

75.   **Yohe, G.** and Schlesinger, M., "The One-Percent Climate Policy", *Hartford Courant,* June 2006.

76.   **Yohe, G.**, Malone, E., Brenkert, A., Schlesinger, M, Mein, H, Xing, X. and Lee, D., "A Synthetic Assessment of the Global Distribution of Vulnerability to Climate Change from the IPCC Perspective that Reflects Exposure and Adaptive Capacity", Palisades, New York: CIESIN (Center for International Earth Science Information Network), 2006.

77.   **Yohe, G.**, "Energy Policy in an Uncertain World Threatened by Dangerous Climate Change", www.lugar.senate.gov/energy, 2006.

**2007**

78.   **Yohe, G.**, "Using Adaptive Capacity to Gain Access to the Decision-intensive Ministries", in *Human-induced Climate Change: An Interdisciplinary Assessment* (Schlesinger, M., ed.), Cambridge: Cambridge University Press, 2007.

79.   **Yohe, G.**, "Lessons for Mitigation from the Foundations of Monetary Policy in the United States", in *Human-induced Climate Change: An Interdisciplinary Assessment* (Schlesinger, M. ed.), Cambridge: Cambridge University Press, 2007.

80.   **Yohe, G.** and Strzepek, K., "Adaptation and Mitigation as Complementary Tools for Reducing the Risk of Climate Impacts", *Mitigation and Adaptation Strategies for Global Change* **12**: 727-739, 2007.

81.   Tol, R., Ebi, K., and **Yohe, G.**, "Infectious Disease, Development and Climate Change: A Scenario Analysis", *Environment and Development Economics* **12**: 687-706, 2007.

82.   √ √ √ Tol, R. and **Yohe, G.**, "The Weakest Link Hypothesis for Adaptive Capacity: An Empirical Test", *Global Environmental Change* **17**: 218-227, 2007.

83. **Yohe, G.** and Tol, R., "The Stern Review: Implications for Climate Change", *Environment,* **40**: 36-41, March 2007.

84. **Yohe, G.**, Tol, R. and Murphy, D., "On Setting Near-term Climate Policy while the Dust Begins to Settle: The Legacy of the *Stern Review"*, *Energy and Environment* **18**: 621-633, 2007.

85. Tol, R.S.J. and **Yohe, G.**, "Infinite Uncertainty, Forgotten Feedbacks, and Cost-Benefit Analysis of Climate Policy", *Climatic Change* **83**: 429-442, 2007.

86. **Yohe, G.**, R.D. Lasco, Q.K. Ahmad, N. Arnell, S.J. Cohen, C. Hope, A.C. Janetos and R.T. Perez, "Perspectives on climate change and sustainability", in *Climate Change 2007: Impacts, Adaptation and Vulnerability. Contribution of Working Group II to the Fourth Assessment Report of the Intergovernmental Panel on Climate Change,* (M.L. Parry, O.F. Canziani, J.P. Palutikof, C.E. Hanson and P.J. van der Linden, eds.), Cambridge: Cambridge University Press, 2007.

87. √ √ Bernstein, L., Bosch, P., Canziani, O., Chen, Z., Christ, R., Davidson, O., Hare, W., Huq, S., Karoly, D., Kattsov, V., Kundzewicz, Z., Liu, J., Lohmann, U., Manning, M., Matsuno, T., Menne, B., Metz, B., Mirza, M., Nicholls, N., Nurse, L., Pachauri, R., Palutikof, J., Parry, M., Qin, D., Ravindranath, N., Reisinger, A., Ren, J., Riahi, K., Rosenzweig, C., Rusticucci, M., Schneider, S., Sokona, Y., Solomon, S., Stott, P., Stouffer, R., Sugiyama, T., Swart, R., Tirpak, D., Vogel, C., and **Yohe, G.**, *Climate Change 2007: Synthesis Report (for the Fourth Assessment Report of the Intergovernmental Panel on Climate Change),* Cambridge: Cambridge University Press, 2007.

88. Janetos, A., Balstad, R., Apt, J., Ardanuy, P, Friedl, R., Goodchild, M., Macauley, M., McBean, G., Skole, D., Welling, L., Wilbanks, T., and **Yohe, G.**, "Earth Science Applications to Societal Benefits", in *Earth Science and Applications from Space: National Imperatives for the Next Decade and Beyond,* Washington, D.C.: National Research Council, The National Academies Press, 2007.

89. **Yohe, G.** and Lasco, R., "Climate and Development Plans must be Combined", SciDev.Net, April 16, 2007.

90. **Yohe, G.**, "A Roadmap for Implementing Adaptation Policy", *Tiempo Climate Newswatch*, April 2007.

91. **Yohe, G.**, "An Issue of Equity – Review of *Fairness in Adaptation to Climate Change*" in *Nature Reports – Climate Change,* (W.N. Adger, J. Paavola, S. Huq and M.J. Mace, eds.) Cambridge, MA: MIT Press, 2007, pp. 319.

92.    **Yohe, G.,** "Thoughts on 'The Social Cost of Carbon: Trends, Outliers and Catastrophes", 2007.
93.    **Yohe, G.,** Burton, I., Huq, S., and Rosegrant, M., "Climate Change: Pro-poor Adaptation, Risk Management, and Mitigation Strategies", 2020 Focus Brief on the World's Poor and Hungry People, Washington, DC: IFPRI, October, 2007.
94.    **Yohe, G.,** "Climate Change in Connecticut", What's New in Science and Technology, *Hartford Courant,* November 1, 2007.
95.    **Yohe, G.,** "Clash: What will Climate Change Cost Us?", *Scientific American,* November 26, 2007.

**2008**

96.    Jones, R. and **Yohe, G.,** "Applying Risk Analytic Techniques to the Integrated Assessment of Climate Policy Benefits", *Integrated Assessment Journal* **8**: 123-149, 2008.
97.    **Yohe, G.** and Tirpak, D., "Summary Report: OECD Global Forum on Sustainable Development: the Economic Benefits of Climate Change Policies (6-7 July 2006)", ENV/EPOC/GSP(2006)11, OECD, Paris, 2006 and forthcoming in *Integrated Assessment Journal* **8**: 1-17, 2008.
98.    Strzepek, K., **Yohe, G.,** Tol, R.J.S., and Rosengrant, M., "The Value of the High Aswan Dam to the Egyptian Economy", *Ecological Economics* **66**: 117-126, 2008.
99.    **Yohe, G.** and Tol, R., "The Stern Review and the Economics of Climate Change: an Editorial Essay", *Climatic Change* **89**:231-240, 2008.
100.   **Yohe, G.,** "Climate Change" in *Solutions for the World's Biggest Problems: Costs and Benefits,* Copenhagen Consensus 2006, Cambridge: Cambridge University Press, 2008.
101.   **Yohe, G.,** "A Reason for Optimism" *Tiempo Climate Newswatch,* January 2008.
102.   **Yohe, G.,** "Inside the Climate Change Panel: Babel of Voices, a Single Conviction", *The InterDependent* **5**: 14, Winter 2007/2008.
103.   **Yohe, G.,** "Climate Change is Real, Compelling and Urgent", *The Guardian,* August 22, 2008.
104.   Lomborg, B. and **Yohe, G.,** "It's Not About Us", *The Guardian,* September 1, 2008.
105.   Keller, K., **Yohe, G.,** Schlesinger, M., "Managing the Risks of Climate Thresholds: Uncertainties and Information Needs", *Climatic Change* **91**: 5-10, 2008.

106. **Yohe, G.** and Tol, R.S.J., "The *Stern Review* and the Economics of Climate Change: An Editorial Essay", *Climatic Change* **89**: 231-240, 2008.

107. **Yohe, G.**, Tol, R.S.J., Richels, R.G., and Blanford, G.J., "Climate Change", Chapter 5 in Lomborg, B. (ed.) for the Copenhagen Consensus 2008, *Global Crises, Global Solutions: Costs and Benefits,* 236-280, 2009.

108. Tol, R.S.J., Richels, R. and **Yohe, G.**, "Future Scenarios for Emissions need Continual Adjustment", *Nature,* **453**: 155, 2008.

109. Moser, S.C., Kasperson, R.E., **Yohe, G.**, and Agyeman, J., "Adaptation to Climate Change in the Northeast: Opportunities, Processes, Constraints", *Mitigation and Adaptation Strategies for Global Change* **13**: 643-659, 2008.

110. **Yohe, G.** and Tol, R.S.J., "Precaution and a Dismal Theorem: Implications for Climate Policy and Climate Research", in *Risk Management in Commodity Markets*, Helyette Geman (ed.), Chichester, UK: John Wiley & Sons Ltd., 2008, pp 91-99, 2008.

111. Keller, K., **Yohe, G.**, and Schlesinger, M., "Managing the Risk of Climate Thresholds: Uncertainties and Information Needs", *Climatic Change* **91**: 5-10, 2008.

112. Keller, K., Tol, R.S.J., Toth, F.L., and **Yohe, G.**, "Abrupt Climate Change near the Poles", *Climatic Change* **91**: 1-4, 2008.

113. **Yohe, G.**, "Characterizing the Value of Reducing Greenhouse Gas Emissions: Creating Benefit Profiles Tracking Diminished Risk", United States Environmental Protection Agency, Washington, DC, 2008.

**2009**

114. Weyant, J., Azar, C., Kainuma, M., Kejun, J., Nakicenovic, N., Shukla, P.R., La Rovere, E. and **Yohe, G.**, "Report of 2.6 Versus 2.9 Watts/m2 RCPP Evaluation Panel", March 31, 2009; submitted to the IPCC Bureau in Istanbul on April 23, 2009.

115. √ Anthoff, D., Tol, R.S.J, and **Yohe, G.**, "Risk Aversion, Time Preference, and the Social Cost of Carbon", *Environmental Research Letters* **4** (2–2): 1–7, 2009.

116. Tol, R.S.J. and **Yohe, G.** "The *Stern Review*: A Deconstruction", *Energy Policy* **37**: 1032–1040, 2009.

117. √ √ √ Smith, J.B., Schneider, S. H., Oppenheimer, M., **Yohe, G.**, Hare, W., Mastrandrea,, M.D., Patwardhan, A., Burton, I., Corfee-Morlot, J., Magadza, C.H.D., Füssel, H-M, Pittock, A.B., Rahman,

A., Suarez, A., and van Ypersele, J-P, "Dangerous Climate Change: An Update of the IPCC Reasons for Concern", *Proceedings of the National Academy of Science* **106**: 4133-4137, March 17, 2009.

118. Anthoff, D., Tol, R.S.J, and **Yohe, G.**, "Discounting for Climate Change", *Economics: The Open-Access, Open-Assessment E-Journal*, **3**: 2009-24, 2009.

119. **Yohe, G.** "Addressing Climate Change through a Risk Management Lens," in *Assessing the Benefits of Avoided Climate Change: Cost Benefit Analysis and Beyond.*, (Gulledge, J., Richardson, L., Adkins, L., and Seidel, S., eds.), *Proceedings of Workshop on Assessing Benefits of Avoided Climate Change, March 16–17, 2009*, Arlington, VA: Pew Center on Global Climate Change, 2009, p. 201–231.

120. **Yohe, G.**, "Toward an Integrated Framework Derived from a Risk-Management Approach to Climate Change", *Climatic Change,* **95**: 325-339, 2009.

121. Collier, W.M., Jacobs, K.R., Saxema, A., Baker-Gallegosi, J., Carroll, M., and **Yohe, G.W.**, "Strengthening Socio-ecological Resilience through Disaster Risk Reduction and Climate Change Adaptation: Identifying Gaps in an Uncertain World", *Environmental Hazards* **8**:171-186, 2009.

122. **Yohe, G.**, "On the Extraordinary Value of 'Committing to Commit' – An Opportunity Not to be Missed", *Climatic Change Letters* **1** published in *Climatic Change* **97**: 285-288, 2009.

123. **Yohe, G.** and Leichenko, R., "Adopting a Risk-Based Approach", in New York City Panel on Climate Change, 2009, *Climate Change Adaptation in New York City: Building a Risk Management Response. (*C. Rosenzweig & W. Solecki, eds.), prepared for use by the New York City Climate Change Adaptation Task Force, *Annals of the New York Academy of Sciences,* New York, NY, 2009, pp. 349. http://onlinelibrary.wiley.com/doi/10.1111/j.1749-6632.2009.05310.x/full.

**2010**

124. **Yohe, G.**, "Addressing Climate Change through a Risk Management Lens – An Overview of Analytic Approaches for Climate Change Based on a Deconstruction of Synthetic Conclusions of the Fourth Assessment Report of the Intergovernmental Panel on Climate Change", in *Assessing the Benefits of Avoided Climate Change: Cost-Benefit Analysis and Beyond.* Proceedings of Workshop on

Assessing the Benefits of Avoided Climate Change, Washington, DC, March 16-17, 2009. Arlington, VA: Pew Center on Global Climate Change, 2010.

125. **Yohe, G.**, "Evaluating Climate Risks in Coastal Zones", in *Issues of the Day: 100 Commentaries on Climate, Energy, the Environment, Transportation and Public Health Policy,* Resources for the Future, 2010.

126.  **Yohe, G.**, "Reasons for Concern" (about Climate Change) in the United States", *Climatic Change* **99**: 295-302, 2010.

127.  √ √ Mastrandrea, M., Field, C., Stocker, T., Edenhofer, O., Ebi, K., Frame, D., Held, H., Kriegler, E., Mach, K., Plattner, G-K., **Yohe, G.**, Zwiers, F., "Guidance Notes for Lead Authors of the IPCC Fifth Assessment Report on Consistent Treatment of Uncertainties", IPCC, 2010.

128. **Yohe, G.,** "The Economics of Climate Change: An Editorial Essay", *Wiley Interdisciplinary Reviews: Climate Change,* 2010.

129.  Blanford, G.J., Richels, R.G., Tol, R.S.J., and **Yohe, G.,** "The Inappropriate Treatment of Climate Change in Copenhagen Consensus 2008", *Climate Change Economics 1:* 135-140, 2010.

130. **Yohe, G.,** "Risk Assessment and Risk Management for Infrastructure Planning and Investment", *The Bridge* **40**(3): 14-21, National Academy of Engineering, 2010.

131. **Yohe, G.,** "Thoughts on *100 volumes of Climatic Change*", *Climatic Change* **100**: 11-14 2010.

132. Rosegrant, M., **Yohe, G.**, Ewing, M., Valmonte-Santos, R., Zhu, T., Burton, I., Huq, S., "Climate Change and Asian Agriculture", *Asian Journal of Agriculture and Development* 7: 1-41, 2010. *Adapting to the Impacts of Climate Change,* Report of the Panel on Adapting to the Impacts of Climate Change, America's Climate Choices, National Research Council, May 19, 2010, available in summary and full text at
https://ageconsearch.umn.edu/bitstream/199082/2/AJAD_2010_7_ 1_3Rosegrant.pdf.

133. **Yohe, G.,** Review of *The Global Deal* by Lord Nicholas Stern, *Journal of Economic Literature,* **48**: 781-786, 2010.

134. **Yohe, G.,** "Risk Management and Climate Change", *Encyclopedia of Climate and Weather* (2nd edition)*,* Oxford: Oxford University Press, 2010.

135. **Yohe, G.** "Economics of Climate Change", *Encyclopedia of Climate and Weather* (2nd edition)*,* Oxford: Oxford University Press, 2010.

136.   **Yohe, G.** with many others, *Climate Stabilization Targets: Emissions, Concentrations, and Impacts Over Decades to Millennia,* Report of the Committee on Climate Stabilization Targets for Atmospheric Greenhouse Gas Concentrations, National Research Council, July 23, 2010.

137.   Bowman, T., Maibach, E., Mann, M., Somerville, R., Seltser, B., Fischoff, B., Gardiner, S., Gould, R., Leiserowitz, and **Yohe, G.,** "Time to Take Action on Climate Communication", *Science* **330**:1044, November 19, 2010.

138.   Wei, D.* and **Yohe, G.,** "The Alternative Energy Race – Career Options for Engineers", American Society of Mechanical Engineers, News Public Policy, November 2, 2010.

139.   **Yohe, G.,** "Thoughts on Addressing Short Term Pressing Needs in a Dynamic and Changing Climate", World Resources Report 2011, December, 2010.

**2011**

140.   **Yohe, G.,** Knee, K. and Kirshen, P., "On the Economics of Coastal Adaptation Solutions in an Uncertain World", *Climatic Change* **106**: 71-92, 2011.

141.   √ Rosenzweig, C., Solecki, W., Gornitz, V., Horton, R., Major, D., **Yohe, G.,** Zimmerman, R., "Developing Coastal Adaptation to Climate Change in the New York City Infrastructure-shed: Process, Approach, Tools, and Strategies", *Climatic Change* **106**: 93-127, 2011.

142.   √ Strzepek, K., **Yohe, G.,** Neumann, J., and Boehlert, B., "Characterizing Changes in Drought Risk for the United States from Climate Change", *Environmental Research Letters* **5** 044012, 2011.

143.   **Yohe, G.,** "Economics: Opportunities from Uncertainties", *Nature Climate Change* **1**: 198-200, June 2011.

144.   **Yohe, G.** and Oppenheimer, M., "Evaluation, Characterization, and Communication of Uncertainty by the Intergovernmental Panel on Climate Change – An Introductory Essay", *Climatic Change* **108**: 629-639, 2011.

## 2012

145.  Iverson, L.R., Matthews, S.N., Prasad, A.M., Peters, M.P., and **Yohe, G.W.**, "Development of Risk Matrices for Evaluating Climatic Change Responses of Forested Habitats", *Climatic Change,* 2012.

146.  **Yohe, G.** and Hope, C., "Some Thoughts on the Value Added from a New Round of Climate Change Damage Estimates," *Climatic Change.*

147.  **Yohe, G., ....** (as one of 39 authors), "1", *Wall Street Journal,* February 1, 2012.

148.  Poulos, H., Bannon, B. Issard, J., Stonebraker. P., Royer, D., **Yohe, G.**, and Chernoff, B., "The World through an Interdisciplinary Lens", *American Association of University Professors,* September, 2012.

149.  **Yohe, G.**, "1", *The Conversation,* October 29, 2012.

150.  **Yohe, G.**, "Republicans Trust Voter Modelling – Why Not Climate Modeling?", *The Conversation,* November 12, 2012.

## 2013

151.  Ebi, K. L and **Yohe, G.**, "Adaptation in First and Second Best Worlds", *SciVerse Science Direct, Environmental Sustainability,* 2013.

152.  Kunreuther, H, Heal, G., Allen, M., Edenhofer, O., Field, C., and **Yohe, G.**, "Risk Management and Climate Change", *Nature Climate Change,* March 2013.

153.  Stone, D., Auffhammer, M., Carey, M., Hansen, G., Huggel, C., Cramer, W., Lobell, D., Molau, U., Solow, A., Tibig, L., and **Yohe, G.**, "The Challenge to Detect and Attribute Effects of Climate Change on Human and Natural Systems", *Climatic Change* **121:** 381-396, August, 2013.

154.  **Yohe, G.** and Hope, C., "Some Thoughts on the Value Added from a New Round of Climate Change Damage Estimates", *Climatic Change* **117:** 451-465, 2013.

155.  **Yohe, G.**, "Climate Change Adaptation – a Risk-Management Approach" in *Handbook of Sustainable Development* (2nd edition) (Atkinson, G., Bietz, S., Neumayer, E. and Agarwala, M., eds.), Cheltenham, UK: Edward Elgar, 2013.

**2014**

156.  Woodhall, C.W., Domke, G.M., Riley, K.L., Oswalt, C.M., Crocker, S.J., and **Yohe, G.**, "A Framework for Assessing Global Change Risks to US Forest Carbon Stocks", *PLO SONE*, 2014.

157.  Anderegg, W., Callaway, E., Boykoff, M.T., **Yohe, G.**, Root, T., "Awareness of Both Type I and II Errors in Climate Science and Assessment", *Bulletin of the American Meteorological Society*, 2014.

158.  **Yohe, G.**, "Economics of Climate Change", *International Encyclopedia of the Social and Behavioral Sciences* (2$^{nd}$ edition; Wright, J, ed.), Oxford, UK: Elsevier Limited, forthcoming in 2014.

159.  Ojima, D.S., Iverson, L.R., Sohngen, B.L., Vose, J.M., Woodhall, C.W., Domke, G.M., Peterson, D.L., Littell, J.S., Matthews, S.N., Prasad, A.M., Peters, M.P., **Yohe, G.W.**, Friggens, M.M., "Risk Assessment" in *Climate Change and United States Forests* (Peterson, D.L., Vose, J.M., and Patel-Weynand, T., eds.), Dordrechi, NE: Springer Publishers, 2014.

160.  Melillo, J, Richmond, T.C., and **Yohe, G.**, and many others, *Third National Climate Assessment,* the White House, United States Global Change Research Program and the National Oceanic and Atmospheric Administration, May 6, 2014 (Overall report, *Highlights*, *Overview*, and Regional Reports available at www.nca2014.globalchange.gov).

161.  Horton, R. and **Yohe, G.**, Easterling, W., Kates, R., Ruth, M., Sussman, E., Whelchel, A., and Wolfe, D., "Northeast Region" in the *Third National Climate Assessment*, the White House, United States Global Change Research Program and the National Oceanic and Atmospheric Administration, May 6, 2014 (part of the Overall report and summarized in *Highlights*, *Overview*, and the Northeast Regional Report).

162.  **Yohe, G.**, "Thoughts on the Context of Adaptation to Climate Change" in *Applied Studies in Climate Adaptation* (Palutikof, Boulter, Barnett and Rissik, eds.), Oxford: Oxford University Press, 2014.

163.  **Yohe, G.**, "Thoughts on the Limits of Adaptation" in *Applied Studies in Climate Adaptation* (Palutikof, J., Boulter, S, and Rissik, D. eds.), Oxford, UK: Wiley-Blackwell, 2014.

164.  **Yohe, G.**, "Some Extending Thoughts on "Thinking Globally and Siting Locally" – Renewable Energy and Biodiversity in a Rapidly Warming World", *Climatic Change* **126**: 7-11, 2014,

https://doi.org/10.1007/s10584-014-1205-1.

165. **Yohe, G.**, "Climate Change Adaptation: A Risk Management Approach", in *Handbook of Sustainable Development,* Atkinson (ed), Cambridge: Cambridge University Press, 2014.

166. Cramer, W., **Yohe, G.**, Auffhammer, M., Huggel, C., Molau, U., Assuncao Faus da Silvia Dias, M., Solow, A., Stone, D., and Tibig, L., "Detection and Attribution of Observed Impacts", in *Climate Change 2014: Impacts, Adaptation and Vulernability. Part A: Global and Sectoral Aspects. Contribution of Working Group II to the Fifth Assessment Report of the IPCC,* Cambridge: Cambridge University Press, 2014.

167. Burkett, V., Suarez, A, Bindi, M., Conde, C., Mukerji, R., Prather, M., Lera St. Clair, A, and **Yohe, G.**, "Point of Departure", Chapter 1 in *Climate Change 2014: Impacts, Adaptation and Vulernability. Part A: Global and Sectoral Aspects. Contribution of Working Group II to the Fifth Assessment Report of the IPCC,* Cambridge: Cambridge University Press, 2014.

## 2015

168. Yohe, G. (with many other members of the NPCC), "NPCC 2015: Building the Knowledge Base for Climate Resiliency*",* New York Panel on Climate Change 2015 Report, (Rosenzweig and Solecki, eds.), *Ann. N.Y Acad. Sci,* **1336:** 1-149.

## 2016

169. Saxena, A., Guneralp, B., Ballis, R., **Yohe, G.**, Chadwick, O., "Evaluating the resilience of forest dependent communities in Central India by combining the sustainable livelihoods framework and cross-scale resilience analysis", *Current Science*: **110**(7), 1195-1206, 2016.

170. Ebi, K., Ziska, L, **Yohe, G.**, "The shape of impacts to come: lessons and opportunities for adaptation from uneven increases in global and regional temperature", *Climatic Change* **139**: 341-349, 2016.

## 2017

171. O'Neill, B.C., Oppenheimer, M., Warren, R., Hallegatte, S., Kopp, R., Portner, H., Scholes, R., Birkmann, J., Foden, W., Licher, R., Mach, K., Marbaiz, P., Mastrandrea, M., Price, J., Takahashi, K., van

Ypersele, J-P., and **Yohe, G.**, "IPCC Reasons for Concern regarding climate change risks", *Nature Climate Change* **7**: 38-37, 2017.

172.   **Yohe, G.**, "A Strong Rebuttal to an Op-Ed by Senator Rand Paul", May 22, 2017, Fox News Opinion.

173.   **Yohe, G.**, "Making Infrastructure Great Again Means Acknowledging Climate Change", Huffington Post, July 12, 2017.

174.   **Yohe, G.**, "Trump's Rejection of the National Climate Report would do More Damage than Exiting the Paris Agreement", *The Conversation*, August 16, 2017.

175.   **Yohe, G.** and Ebi, K., "Doctors Must Respond to Changes in the Political Climate of Climate Change", STATnews First Opinion, August 22, 2017.

176.   **Yohe, G.** "Why is this missing? @realDonaldTrump #prepare Get prepared! The worst is yet to come", World Government Research Network.

177.   **Yohe, G.**, "Characterizing Transient Temperature Trajectories for Assessing the Value of Achieving Alternative Temperature Targets", *Climatic Change*, **145**: 469-479, 2017.

178.   **Yohe, G.**, "The climate science report Trump hoped to ignore will resonate outside of Washington, DC", *The Conversation,* November 8, 2017.

**2018**

179.   Travis, W., Smith, J., **Yohe, G.**, "Moving toward 1.5°C of warming: implications for climate adaptation strategies", *Current Opinion in Environmental Sustainability,* Volume 31, April 2018, pg 146-152, https://www.sciencedirect.com/journal/current-opinion-in-environmental-sustainability/vol/31/suppl/C .

180.   **Yohe, G.** with many others on the Committee to Review the Draft Fourth National Climate Assessment, *Review of the Draft Fourth National Climate Assessment,* National Academies of Sciences, March 12, 2018, Washington, DC, https://NAS.edu .

181.   **Yohe, G.**, "Trump's immoral response to the climate report", Hartford Courant (picked up across the Tribune network), November 28, 2018; https://www.courant.com/opinion/op-ed/hc-op-yohe-trump-immoral-climate-response-20181127-story.html.

182.   **Yohe, G.** and Mann, M., "People are already dying by the thousands because of climate change", Huffington Post, December 11, 2018, https://www.huffingtonpost.com/entry/opinion-climate-change-deaths_us_5c101e14e4b0ac5371799b1c .

183. Richels, R., **Yohe, G.,** and Jacoby, H., 2018, "The Naked Truth About US Climate Policy (December 9, 2018). Available at SSRN: www.https://ssrn.com/abstract=3298322 or www.http://dx.doi.org/10.2139/ssrn.3298322.

**2019**

184. Xi, L. and **Yohe, G.,** "Adaptation in an uncertain world – detection and attribution of climate change trends and extreme possibilities" in *Planning for climate change hazards"* (Smith, J. and Pfeffer, W.T., eds), Oxford: Oxford University Press, 2019. https://www.oxfordhandbooks.com/view/10.1093/oxfordhb/978019 0455811.001.0001/oxfordhb-9780190455811-e-60?rskey=UMtkrF&result=1

185. Blake, R., Klaus, J., **Yohe, G.,** Zimmerman, R., Manley, D., Rosenzweig, C., and Solecki, W., "Chapter 8: Indicators and Monitoring", in *New York City Panel on Climate Change 2019 Report, Annals of the New York Academy of Sciences,* New York Academy of Sciences, 2019. doi: 10.1111/nyas at: https://nyaspubs.onlinelibrary.wiley.com/toc/17496632/2019/1439/ 1.

186. **Yohe, G.,** Richels, R, and Jacoby,H., 2019, "Don't let the 'Green New Deal' hijack climate policy", https://www.econ**news**.site.wesleyan.edu/2019/03/02/**yohe...green-new-deal**-in-courant.

187. Moss, R.H., Avery, S., Baja, K., Burkett, V., Chischilly, A.M., Dell, J., Fleming, P.A., Geil, K., Jacobs, K., Jones, A., Knowlton, K., Koh, J., M. C. Lemos, M.C., Melillo, J., Pandya, R., Ricnmond, T.C., l. Scarlett, L., Snyder, J., Stults, M., Waple, A., Whitehead, J., Zarrilli, D., Fox, J., Ganguly, A., Joppa, L., Julius, S., Kirshen, P., Kreutter, R., Mcgovern, A., Meyer, R., Neumann, J., Solecki, W., Smith, J., Tissot, P., **Yohe, G.,** and Zimmerman, R., 2019, "A framework for sustained climate assessment in the United States", *Bulletin of the American Meterological Society,* **100** (5): 897–907, https://doi.org/10.1175/BAMS-D-19-0130.1

188. Richels, R., **Yohe, G.,** and Jacoby, H., 2019, "Carbon dioxide is not our grandparents' pollutant", *Index of Earth and Planetary Sciences,* https://www.thescientificnews.com/carbon-dioxide-co2-is-not-our-grandparents-pollutant-by-richard-richels-gary-yohe-and-henry-jacoby/.

189.    **Yohe, G.,** 2019, "A review of 'Discerning Experts', *Climatic Change,* 155(3), 295-309,

190.    **Yohe, G.,** The economic cost of devastating hurricanes and other extreme weather events is even worse than we thought", *The Conversation,* May 31, 2019, https://theconversation.com/the-economic-cost-of-devastating-hurricanes-and-other-extreme-weather-events-is-even-worse-than-we-thought-108315?utm_medium=ampemail&utm_source=email.

191.    **Yohe, G,** "A $1trillion economic blow? The cost of extreme weather in the U.S. is worse than we thought", *Washington Post*, June 9th, 2019, https://www.washingtonpost.com/weather/2019/06/07/trillion-economic-blow-cost-extreme-weather-us-is-worse-than-we-thought/?utm_term=.e72223eae058.

192.    Richels, R., Jacoby, H., **Yohe, G.,** "Advice on climate policy for the 2020 presidential candidates", *The Hill,* July 18, 2019, https://**thehill.com**/opinion/energy-environment/453755-advice-on-climate-policy-for-the 2020 presidential candidatese.

193.    **Yohe, G.** and Mann, M., "Two kinds of climate, one thing in common", *The Hartford Courant*, August 25, 2019, https://www.courant.com/opinion/op-ed/hc-op-yohe-mann-climate-change-0825-20190825-cblr6vhvuzaj7po7jmhpjegqme-story.html.

194.    **Yohe, G.,** "Why damage estimates for hurricanes like Dorian won't capture the full cost of climate change-fueled disasters", *The Conversation,* September 3, 2019, http://theconversation.com/why-damage-estimates-for-hurricanes-like-dorian-wont-capture-the-full-cost-of-climate-change-fueled-disasters-122910.

195.    Richels, R., Jacoby, H. and **Yohe, G.,** "A tragic misperception about climate change", *The Hill*, September 5, 2019, https://thehill.com/opinion/energy-environment/459980-a-tragic-misperception-about-climate-change?rnd=1567632529.

196.    **Yohe, G.,** Richels, R. and Jacoby, H., "Adapt, abate, or suffer – lessons from Hurricane Dorian", *The Globe Post*, October 23, 2019, https://theglobepost.com/2019/10/22/lessons-hurricane-dorian/.

197.    Richels, R., **Yohe, G.,** and Jacoby, H., "Who is holding up the war on global warming? You may be surprised", *The Hill*, November 2, 2019, https://thehill.com/opinion/energy-environment/468677-who-is-holding-up-the-war-on-global-warming-you-may-be-surprised?rnd=1572714739

198.    Richels, R., Yohe, G. and Jacoby, H., "There is no plan B on climate change, *The Hill,* December 20, 2019,

https://thehill.com/opinion/energy-environment/474994-there-is-no-plan-b-on-climate-change?rnd=1576622411.

**2020**

199. Neumann, J., Willwerth, J., Martinich, J., McFarland, J., Saofim, M. and **Yohe, G.**, "Climate damage functions for estimating the economic impacts of climate change in the United States", 2020, *Review of Environmental Economics and Policy.* Accessed: https://academic.oup.com/reep/article-abstract/14/1/25/5699606; and
https://academic.oup.com/reep/article/14/1/25/5699606?guestAccessKey=c4e1a2ce-75d3-407d-bcf9-a4d623cfca5c.

200. **Yohe, G.**, "Climate Action and Policy – Parallels with COVID-19", Earth Day 2020, *Springer Nature*,
https://www.springernature.com/gp/researchers/sdg-programme/earth-day

201. Jacoby, H., Richels, R., **Yohe, G.,** and Santer, B., "Can a pandemic aid the fight against global warming?", *The Hill,* May 16, 2020,
https://thehill.com/opinion/energy-environment/498145-can-a-pandemic-aid-the-fight-against-global-warming?rnd=1589658368.

202. Richels, R., Jacoby, H, **Yohe, G.,** and Santer, B., "We cannot ignore the links between COVID-19 and the warming planet", *The Hill,* May 27, 2020,
https://thehill.com/opinion/energy-environment/499604-we-cannot-ignore-the-links-between-covid-19-and-the-warming-planet?rnd=1590527443.

203. **Yohe, G.,** "The coronavirus is showing the cracks in the foundation of American society", *The Hartford Courant,* June 27, 2020,
https://www.courant.com/opinion/op-ed/hc-op-yohe-revelations-of-coronavirus-0627-20200627-uhmas4jd5neszh5tbgv6ch26he-story.html.

204. **Yohe, G.**, Santer, B., Jacoby, H., Richels, R., "Counterfactual experiments are crucial but easy to misunderstand", *Scientific American*, July 9, 2020,
https://www.scientificamerican.com/article/counterfactual-experiments-are-crucial-but-easy-to-misunderstand/.

205. Richels, R., Jacoby, H., **Yohe, G.,** and Santer, B., "The Trump administration cooks the climate change numbers once again", *The Hill*, July 19, 2020,

https://thehill.com/opinion/energy-environment/507929-the-trump-administration-cooks-the-climate-change-numbers-once#bottom-story-socials and https://apple.news/A19pGk94PSbiGqdPdul4HlA.

206. **Yohe, G.** Willwerth, J., Neumann, J., and Kerrich, Z., "What the future might hold: Regional transient sectoral damages for the United States – Estimates and maps in an exhibition", *Climate Change Economics* **11**(4): 204001-24, https://doi.org/10.1142/S2010007820400023

207. **Yohe, G.**, Santer, B., Jacoby, H., and Richels, R., "Five science questions that ought to be asked at the debate", *Yale Climate Communications,* September 4, 2020, https://yaleclimateconnections.org/2020/09/five-science-questions-that-ought-to-be-asked-at-the-debate/

208. **Yohe, G.**, Santer, B., Jacoby, H., and Richels, R., "Key messages about climate change: an introduction to a series", *Yale Climate Connections,* September 11, 2020, https://yaleclimateconnections.org/2020/09/key-messages-about-climate-change-an-introduction-to-a-series/

209. Jacoby, H., **Yohe, G.**, Richels, R., "Evidence shows troubling warming of the planet, *Yale Climate Connections,* September 18, 2020, https://yaleclimateconnections.org/2020/09/evidence-shows-the-planet-warming-on-average-at-an-increasing-rate/

210. **Yohe, G.**, Jacoby, H., Richels, R., "The evidence is compelling on human activity as the principle cause of global warming", *Yale Climate Connections,* September 25, 2020, https://yaleclimateconnections.org/2020/09/the-evidence-is-compelling-on-human-activity-as-the-principal-cause-of-global-warming/

211. Richels, R., Jacoby, H., and **Yohe, G.**, "Extreme events 'presage worse to come' in a warming climate", *Yale Climate Connections,* October 2, 2020, https://yaleclimateconnections.org/2020/10/extreme-events-presage-worse-to-come-in-a-warming-climate/

212. **Yohe, G.**, "Climate change is getting worse, and it's harder to predict", *The Hartford Courant,* October 4, 2020, https://www.courant.com/opinion/op-ed/hc-op-yohe-climate-pendulum-1004-20201004-5xrmc4vkdfhv7nehn7zickakke-story.html

213. **Yohe, G.**, Willwerth, J., Neumann, J., and Kerrich, Z., "What the future might hold: Distributions of regional sectoral damages for the

United States – Estimates and maps in an exhibition", *Climate Change Economics*, **11**: 4, https://www.doi:10.1142/S2010007820400023.

214.  **Yohe, G.**, Jacoby, H., and Richels, R., "Multiple extreme climate events can combine to produce catastrophic damages", *Yale Climate Connections,* October 9, 2020, https://yaleclimateconnections.org/2020/10/multiple-extreme-climate-events-can-combine-to-produce-catastrophic-damages/

215.  Jacoby, H., **Yohe, G.**, and Richels, R., "Vigorous action needed, and soon, on climate change", *Yale Climate Connections,* October 16, 2020,   https://yaleclimateconnections.org/2020/10/vigorous-action-needed-and-soon-on-climate-change/

216.  Richels, R., Jacoby, H., and **Yohe, G.**, "Rejoining the fight against climate change is in the U.S. national interest", *Yale Climate Connections,* October 23, 2020, https://yaleclimateconnections.org/2020/10/rejoining-the-global-fight-against-climate-change-in-the-u-s-national-interest/

217.  **Yohe, G.**, Jacoby, H., Santer, B., and Richels, R., , "Inaction on the climate threat is NOT an option", *Yale Climate Connections,* October 30, 2020 https://yaleclimateconnections.org/2020/10/inaction-on-the-climate-threat-is-not-an-option/

**2021**

218.  **Yohe, G.**, Jacoby, H., Richels, R., and Santer, B., "Early next step: Add risk management to National Climate Assessment", *Yale Climate Connections,* 2021 https://yaleclimateconnections.org/2021/01/commentary-early-next-step-add-risk-management-to-national-climate-assessment/

219.  **Yohe, G.**, Jacoby, H., Santer, B., and Richels, R., "Deadlines loom for Capitol Hill action on Trump-era climate issues", *Yale Climate Connections,* January 2021, https://yaleclimateconnections.org/2021/01/commentary-deadlines-loom-for-capitol-hill-action-on-trump-era-climate-issues/

220.  Jacoby, H., **Yohe, G.**, Santer, B., and Richels, R., "Biden channels FDR on STEM policy", *Scientific American,* February 18, 2021, https://www.scientificamerican.com/article/biden-channels-fdr-on-stem-policy/?previewid=5C22384B-0948-422D-B7F0AE60D40A5F64

221.  **Yohe, G.,** Jacoby, H., Santer, B, and Richels, R., "Biden's executive orders have broad public support", *Yale Climate Connections,* March 8, 2021, https://yaleclimateconnections.org/2021/03/bidens-executive-orders-on-climate-have-broad-public-support/

222.  **Yohe, G.,** Richels, R., and Jacoby, H., "Communicating the essentials of climate risk", Chapter 23 in *Lectures in Climate Change - Our warming planet: climate change impacts and adaptation,* Cambridge: Cambridge University Press, 2021.

223.  **Yohe, G.,** "How to think about climate change responses: on organizing one's thoughts" in *Handbook of climate change mitigation and adaptation* (M. Lackner et al eds), Springer Nature, Springer Science+Business Media, LLC, 2021.
https://link.springer.com/content/pdf/10.1007%2F978-1-4614-6431-0_102-1.pdf and
https://doi.org/10.1007/978-1-4614-6431-0_102-1.

224.  Richels, R., Jacoby, H., Santer, B., and **Yohe, G.,** "The choice is clear: fair climate policy or no climate policy", *Yale Climate Connections,* March 22, 2021,
https://yaleclimateconnections.org/2021/03/commentary-the-choice-is-clear-fair-climate-policy-or-no-climate-policy/

225.  **Yohe, G.,** "Review of climate change is an unjust war – And so a war on climate risk is 'just'", *Academia,* April 11, 2021,
https://www.academia.edu/s/ebe2172b1d?source=ai_email

226.  **Yohe, G.** and E. Rignot, "With seas rising stalled research budgets must also rise" *Yale Climate Connections,* April 27, 2021
https://yaleclimateconnections.org/2021/04/with-seas-rising-stalled-research-budgets-must-also-rise/

227.  **Yohe, G.,** "A new book manages to get science badly wrong", *Scientific American,* May 13, 2021
https://www.scientificamerican.com/article/a-new-book-manages-to-get-climate-science-badly-wrong/.

228.  **Yohe, G.,** "On the value of conducting and communicating counterfactual exercises: Lessons from epidemiology and climate science", in *Environmental Change and Sustainability,* London: InTech, 2021, http://mts.intechopen.com/articles/show/title/on-the-value-of-conducting-and-communicating-counterfactual-exercise-lessons-from-epidemiology-and-c

229.  **Yohe, G.,** "Investing in infrastructure is investing in anything that amplifies the productivity of privately held physical capital", *Academia Letters*, 2021, Article 1867.
https://doi.org/10.20935/AL1867

230.  **Yohe, G.,** "This is how extreme weather events and climate change are connected, *The Hill,* July 26, 2021, https://thehill.com/opinion/energy-environment/564933-this-is-how-extreme-weather-events-and-climate-change-are

231.  **Yohe, G.,** "Climate change and COVID-19: Understanding existential threats", *The Hill,* July 30, 2021, https://thehill.com/opinion/energy-environment/565698-climate-change-and-covid-19-understanding-existential-threats

232.  **Yohe, G.,** "'Never before" (NB4) extreme weather events and near-missess", *Yale Climate Connections,* September 9, 2021, https://yaleclimateconnections.org/2021/09/never-before-nb4-extreme-weather-events-and-near-misses/

233.  **Yohe, G.,** "Review of "Unsettled - What Climate Science Tells Us, What it Doesn't, and Why it Matters" by Steven Koonin, *Energy Journal* Volume 43, Number 3, https://www.iaee.org/energyjournal/issue/3850 https://www.iaee.org/en/publications/init2.aspx?id=0 .

234.  **Yohe, G.,** "On interpreting "Code Red" and "Existential Threats", *World Government Research Network,* November 9, 2021,

235.  Smith, J. and **Yohe, G.,** 2021, "Implement new climate commitments — then go beyond to constrain warming", *The Hill, November 17, 2021,* https://thehill.com/opinion/energy-environment/581885-implement-new-climate-commitments-then-go-beyond-to-constrain

**2022**

236.  Richels, R., Jacoby, H., Santer, B., and **Yohe, G.,** "The 1.5 degree goal: Beware of unintended consequence", *Yale Climate Connections,* January 5, 2022, https://yaleclimateconnections.org/2022/01/the-1-5-degrees-goal-beware-of-unintended-consequences/

237.  **Yohe, G.,** 2022, "A Nobel Prize in Physics that we can all understand", *World Financial Review,* July/August, https://worldfinancialreview.com/a-nobel-prize-in-physics-that-we-can-all-understand/

238.  **Yohe, G.,** "Energy transformation can strengthen democracy and help fight climate change", *Yale Climate Connections,* April 1, 2022, https://yaleclimateconnections.org/2022/04/energy-transformation-can-strengthen-democracy-and-help-fight-climate-change/

239.     **Yohe, G.,** "How to read the latest science report from the IPCC",
         *The Hill*, April 7, 2022,
         https://thehill.com/opinion/energy-environment/3261801-how-to-
         read-the-latest-science-report-from-the-ipcc/
240.     Jacoby, H, Santer, B., **Yohe, G.,** and Richels, R., Fighting climate
         change in a fragmented world, *The Hill,* May 7, 2022,
         https://thehill.com/opinion/energy-environment/3479916-fighting-
         climate-change-in-a-fragmented-world/
241.     Richels, R., Santer, B., Jacoby, H., and **Yohe, G.,** "A durable U.S.
         climate strategy or a house of cards", *Yale Climate Connections,*
         June 6, 2022, https://yaleclimateconnections.org/2022/06/a-durable-
         u-s-climate-strategy-or-a-house-of-cards/
242.     Stuart, B. and **Yohe, G.,** "Could steam powered cars decrease CO2
         in the atmosphere?", *The Conversation,* June 13, 2022,
         https://theconversation.com/could-steam-powered-cars-decrease-
         the-co2-in-the-atmosphere-182917 and
         https://theconversation.com/bisakah-mobil-bertenaga-uap-mengurangi-
         co2-di-atmosfer-186260
243.     **Yohe, G.,** "Supreme court EPA climate ruling: what did Congress
         intend with the Clear Cir Act?", *The Hill,* July 2, 2022,
         https://thehill.com/opinion/energy-environment/3544685-supreme-court-
         epa-climate-ruling-what-did-congress-intend-with-clean-air-act/
244.     **Yohe, G.,** "Celebrate climate action but do not let your guard down",
         *The Hill,* August 15, 2022, https://thehill.com/opinion/energy-
         environment/3602772-celebrate-climate-action-but-do-not-let-your-
         guard-down/
245.     **Yohe, G.,** "No, the IRA is not a carbon tax but it would be cheaper
         if it were", *The Hill,* August 25, 2022,
         https://thehill.com/opinion/energy-environment/3614555-no-the-
         ira-is-not-a-carbon-tax-but-it-would-be-cheaper-if-it-were/
246.     **Yohe, G.,** "No, the IRA is not a carbon tax but it would be cheaper
         if it were", *The Hill,* August 25, 2022,
         https://thehill.com/opinion/energy-environment/3614555-no-the-
         ira-is-not-a-carbon-tax-but-it-would-be-cheaper-if-it-were/
247.     **Yohe, G.,** "Antarctica's melting glaciers are no false alarm", *The
         Hill,* October 9, 2022,
         https://thehill.com/opinion/energy-environment/3678506-
         antarcticas-melting-glaciers-are-no-false-alarm/
248.     **Yohe, G.,** "COP27 in Egypt should be the death of climate change
         denial", *CGTN*, November 11, 2022,

https://news.cgtn.com/news/2022-11-11/COP27-in-Egypt-should-be-the-death-of-climate-change-denial-1eRNjRxkTJK/index.html.

249. **Yohe, G.,** "GOP-controlled House: Children playing poorly in the climate change sandbox", *The Hill,* November 22, 2022, https://thehill.com/opinion/energy-environment/3746950-gop-controlled-house-children-playing-poorly-in-the-climate-change-sandbox/

250. **Yohe, G.,** "What's worse than the worst case scenario for climate change?", *Newsweek,* December 28, 2022, https://www.newsweek.com/whats-worse-worst-case-scenario-climate-change-opinion-1769184

**2023**

251. **Yohe, G.,** "What does it mean that (once rare) atmospheric rivers and bomb cyclones are becoming more frequent", *The Hill,* January 7, 2023, https://thehill.com/opinion/energy-environment/3803900-what-does-it-mean-that-once-rare-atmospheric-rivers-and-bomb-cyclones-are-becoming-more-frequent/

252. **Yohe, G.,** January 2023 "Risks and costs of climate change" in *Climate science and law curriculum,,* Climate Judiciary Project, Environmental Law Institute, Washington, D.C, https://www.eli.org/sites/default/files/files-pdf/Risks%20and%20Costs%20of%20Climate%20Change_full%20report%20formatted.pdf

253. **Yohe, G.,** "The WFE's leadership on the environment is a step in the right direction", *China global television news,* January 22, 2023, https://news.cgtn.com/news/2023-01-21/The-WEF-s-leadership-on-environment-is-a-step-in-right-direction-1gM1STm7pT2/index.html.

**Other Professional Activities:**

1. Member, Steering Committee and Executive Committee for the Human Dimensions of Global Environmental Change Research, International Social Science Council (ISSC), 1993 – 1994.

2. Lead Author, chapters 2 (Methods and Tools), 18 (Adapting to Climate Change in the Context of Sustainable Development and Equity) and 19 (Synthesis) for the *Third Assessment Report of Working Group II of the Intergovernmental Panel on Climate Change*, and chapter 1 (Setting the Stage: Climate Change and

Sustainable Development) for *Third Assessment Report of Working Group III of the Intergovernmental Panel on Climate Change,* 1998 – 2001. Consult http://ipcc.ch.

3.  Convening Lead Author, chapter 4, "Recognizing Uncertainty in Assessing Responses", Report of the Policy Responses Working Group, Millennium Ecosystem Assessment, Island Press, 2005. Consult http://www.millenniumassessment.org/en/index.aspx.

4.  Member, Earth Science Applications and Societal Objectives Panel of the Earth Science and Applications from Space Panel of the Space Studies Board of the National Research Council of the National Academies, 2005 – 2006.

5.  Member, Board of Editors (with H.J. Schellnhuber, W. Cramer, N. Nakicenovic, and T. Wigley), for *Avoiding Dangerous Climate Change,* Cambridge University Press, 2005 – the Proceedings of the Symposium on Avoiding Dangerous Climate Change hosted by Tony Blair in Exeter, UK in February of 2005. Consult http://www.cambridge.org/catalogue.

6.  Chair of a Global Forum on the Economic Benefits of Climate Change Policies hosted by the Organization for Economic Cooperation and Development, Paris, July 6-7, 2006; for papers and presentations, consult htpp://www.oecd.org/env/cc/benefitsforum2006.

7.  Member, Climate Change Science Program Product Development Advisory Committee for the United States Department of Energy by the Secretary of Energy. 2006 – 2014.

8.  Member, Council, Connecticut Academy of Science and Engineering, 2001-2007.

9.  Convening Lead Author, chapter 20 in the contribution of Working Group II to the *Fourth Assessment Report of the Intergovernmental Panel on Climate Change* – "Perspectives on Climate Change and Sustainability", 2004 – 2007. Consult http://ipcc.ch.

10. Member, Core Writing Team for the Synthesis Report of the Fourth Assessment Report of the Intergovernmental Panel on Climate Change, 2005-2007. Consult http://ipcc.ch.

11. Testified before the Senate Foreign Relations Committee on the "Hidden (climate change) Cost of Oil" on March 30, 2006, the Senate Energy Committee on the *Stern Review* on February 14, 2007, and the Senate Banking Committee on "Material Risk from Climate Change and Climate Policy" on October 31, 2007.

12. Member, Committee on the Human Dimensions of Global Change of the National Academy of Science, 2007 – present.

13.  Member, Adaptation Subcommittee of the Governor's (CT) Steering Committee on Climate Change, 2008-2010.
14.  Member, National Research Council Committee on America's Climate Choices: Panel on Adapting to the Impacts of Climate Change, 2008-2011.
15.  Member, National Research Council Panel on Addressing the Challenges of Climate Change through the Behavioral and Social Sciences, 2009-2010.
16.  Member, The Sustainability Leadership Council of the Green Education Foundation, June 2009 - present; see www.greeneducationfoundation.org.
17.  Member, National Research Council Committee on Stabilization Targets for Atmospheric Greenhouse Gas Concentrations, 2009-2010.
18.  Participated with our Chair as a representative of the Adaptation Panel in the prepublication briefings to the United States House of Representatives, the United States Senate, the National Oceanographic and Atmospheric Administration, the White House Office of Science and Technology Policy and the White House Council on Environmental Quality, May 18-19, 2010 in Washington, DC.
19.  Member, Planning and Coordinating Committee for the National Adaptation Summit hosted in Washington, DC by the White House Office of Science and Technology Policy, the White House Council for Environmental Quality and the National Oceanographic and Atmospheric Administration, May 25-27, 2010, 2009-2010.
20.  Co-editor-in Chief, *Climatic Change* with Michael Oppenheimer, August 2010-present.
21.  Member, Climate Science Rapid Response Team (SCRRT) organized under the American Geophysical Union, November 2010-present.
22.  Member, Science Steering Group (SSG) for the IPCC Expert Meeting on Economic Analysis, Costing Methods, and Ethics that took place 22-24 June 2011 in Peru.
23.  Lead Author for chapters 1 and 18 for the contribution of Working Group II to the *Fifth Assessment Report of the Intergovernmental Panel on Climate Change*, 2012-2014.
24.  Vice Chair, National Climate Assessment Development and Advisory Committee for the White House, April 2011-August 2014. The Assessment was "rolled out" by the White House on May 6, 2014.

25.   Participated in Briefings for the Congress (Senate and then House of Representatives) upon the release of the National Climate Assessment, May 7, 2014).

26.   Convening Lead Author, Northeast Regional contribution to the National Climate Assessment.

27.   Member, Committee to Advise the United States Global Change Research Program, National Research Council and National Academies of Science, July 2011-2015.

28.   Member of the Board on Environmental Change and Society (BECS), National Research Council and National Academies, January 2012 – January 2014.

29.   Member, Nominating Committee of the American Economic Association, January 2014 – April 2014.

30.   Member, Expert Review Panel for "Risky Business" organized by Michael Bloomberg and a committee of authors that includes Hank Paulson, Robert Rubin, George Schulz, Olympia Snow, and others for the release of an assessment focusing on financial risk from climate Change, January 2014 until the present; release expected in June of 2014.

31.   Member, National Academy of Sciences, Engineering and Medicine Panel on the 2017-2027 Decadal Survey for Earth Science and Applications from Space; April 2016-present.

32.   Member, New York (City) Panel on Climate Change, August 2008 – present.

33.   Appointed by the Chair of the National Research Council to the Committee for the Review of the Draft Fourth National Climate Assessment, November 2017-July 2018.

34.   Participating Member, Science Defenders, January 1, 2019.

35.   Contributing Author to "A Civil Society for Conducting Applied Climate Assessments: Collaborations and Knowledge for Confronting Climate Risk", *Report of an Independent Advisory Committee on Applied Climate Assessment, Weather, Climate and Society* as well as the *Bulletin of the American Meteorological Society,* February 2019.

36.   Reviewer for the IPCC Scholarship Programme, March-April 25, 2019.

37.   Reviewer, chapter 16 of the contribution of Working Group II to the *Sixth IPCC Assessment.*

38.   Featured in April of 2019 on the Coexist Blog of the College of the Environment at Wesleyan; see www. http://bit.ly/2LbM1YS.

39.     Member, SpringerNature United Nations Sustainable Development Goals Focus Group, Climate Action #13, July 2020 – present.

40.     Expert reviewer, chapter 26 of the Contribution of Working Group II to the Sixth Assessment Report of the Intergovernmental Panel on Climate Change. December 2020-February 2021.

41.     Contributing scientist, Climate Power 2020, May 2020-November 2020.

42.     Appointed member, US National Academies Review Panel for the Sixth US National Climate Assessment, October 2022-February 2023.